Lecture Notes in Artificial Inte

Edited by R. Goebel, J. Siekmann, and W. Wahlster

Subseries of Lecture Notes in Computer Science

Petra Perner   Ovidio Salvetti (Eds.)

Advances in Mass Data Analysis of Images and Signals in Medicine, Biotechnology, Chemistry and Food Industry

Third International Conference, MDA 2008
Leipzig, Germany, July 14, 2008
Proceedings

 Springer

Series Editors

Randy Goebel, University of Alberta, Edmonton, Canada
Jörg Siekmann, University of Saarland, Saarbrücken, Germany
Wolfgang Wahlster, DFKI and University of Saarland, Saarbrücken, Germany

Volume Editors

Petra Perner
IBaI - Institute of Computer Vision
and Applied Computer Sciences
Arno-Nitzsche-Str. 43, 04277 Leipzig, Germany
E-mail: pperner@ibai-institut.de

Ovidio Salvetti
Istituto di Scienza e Tecnologie dell'Informazione (ISTI)
Area della Ricerca CNR di Pisa
Via G. Moruzzi 1, 56124 Pisa, Italy
E-mail: ovidio.salvetti@isti.cnr.it

Library of Congress Control Number: Applied for

CR Subject Classification (1998): H.2.8, I.4.6, J.3

LNCS Sublibrary: SL 7 – Artificial Intelligence

ISSN   0302-9743
ISBN-10   3-540-70714-X Springer Berlin Heidelberg New York
ISBN-13   978-3-540-70714-1 Springer Berlin Heidelberg New York

Springer is a part of Springer Science+Business Media

springer.com

© Springer-Verlag Berlin Heidelberg 2008
Printed in Germany

Typesetting: Camera-ready by author, data conversion by Scientific Publishing Services, Chennai, India
Printed on acid-free paper   SPIN: 12441549   06/3180   5 4 3 2 1 0

Preface

The automatic analysis of signals and images together with the characterization and elaboration of their representation features is still a challenging activity in many relevant scientific and hi-tech fields such as medicine, biotechnology, and chemistry. Multidimensional and multisource signal processing can generate a number of information patterns which can be useful to increase the knowledge of several domains for solving complex problems. Furthermore, advanced signal and image manipulation allows relating specific application problems into pattern recognition problems, often implying also the development of KDD and other computational intelligence procedures.

Nevertheless, the amount of data produced by sensors and equipments used in biomedicine, biotechnology and chemistry is usually quite huge and structured, thus strongly pushing the need of investigating advanced models and efficient computational algorithms for automating mass analysis procedures. Accordingly, signal and image understanding approaches able to generate automatically expected outputs become more and more essential, including novel conceptual approaches and system architectures.

The purpose of this third edition of the International Conference on Mass Data Analysis of Signals and Images in Medicine, Biotechnology, Chemistry and Food Industry (MDA 2008; www.mda-signals.de) was to present the broad and growing scientific evidence linking mass data analysis with challenging problems in medicine, biotechnology and chemistry. Scientific and engineering experts convened at the workshop to present the current understanding of image and signal processing and interpretation methods useful for facing various medical and biological problems and exploring the applicability and effectiveness of advanced techniques as solutions.

The primary goal of the conference was to disseminate this knowledge to a multidisciplinary community and encourage cooperative proactive collaboration in all the interested fields.

We were pleased to see that the idea of the conference was taken up by a growing number of researchers and that we could start to bundle the activities in this area.

We appreciate the help and understanding of the editorial staff at Springer, and in particular Alfred Hofmann, who supported the publication of these proceedings in the LNAI series.

Last, but not least, we wish to thank all the speakers and participants who contributed to the success of the conference.

The next International Conferences on Mass Data Analysis of Signals and Images (www.mda-signals.de) will be held in July 2009. We are looking forward to your submissions.

July 2008

Petra Perner
Ovidio Salvetti

International Conferences on Mass Data Analysis of Signals and Images in Medicine, Biotechnology, Chemistry, and Food Industry, MDA 2008

July 14, 2008, Leipzig, Germany

Organization

Institute of Computer Vision and applied Computer Sciences, IBaI, Germany

Chairs

Petra Perner   IBaI, Germany
Ovidio Salvetti   CNR-ISTI, Italy

Committee

Walter Arnold	Fraunhofer Institute of Non-destructive Testing, Germany
Ewert Bengsston	University of Uppsala, Sweden
Valentin Brimkov	Buffalo State College, USA
Hans du Buf	University of Algarve, Portugal
Eugenio Fava	Max Planck Institute of Molecular Cell Biology & Genetics, Germany
Maria Frucci	Istituto di Cibernetica, CNR, Italy
Igor Gurevich	Academy of Science, Russia
Thomas Günther	JenaBios GmbH, Germany
Giulio Iannello	University Campus Bio-Medico of Rome, Italy
Xiaoyi Jiang	University of Muenster, Germany
Montse Pardas	Universitat Politècnica de Catalunya, Spain
Thang Viet Pham	OncoProteomics Laboratory - VUmc, The Netherlands
Gabriella Saniti di Baja	Istituto di Cibernetica, CNR, Italy
Arnold Smeulders	University of Amsterdam, The Netherlands
Tuan Pham	James Cook University, Australia
Julie Wilson York	Structural Biology Laboratory, UK

Aim of Conference

The automatic analysis of images and signals in medicine, biotechnology, and chemistry is a challenging and demanding field.

Signal-producing procedures by microscopes, spectrometers and other sensors have found their way into wide fields of medicine, biotechnology, economy and environmental analysis. With this arises the problem of the automatic mass analysis of signal information. Signal-interpreting systems which generate automatically the desired target statements from the signals are therefore of compelling necessity. The continuation of mass analyses on the basis of the classical procedures leads to investments of proportions that are not feasible. New procedures and system architectures are therefore required.

Scope of Conference

The scope of the International Conference on Mass Data Analysis of Images and Signals in Medicine, Biotechnology, Chemistry and Food Industry (www.mda-signals.de) is to bring together researcher, practitioners and industry people who deal with mass analysis of images and signals to present and discuss recent research in these fields.

The goals of this workshop are to:

- Provide a forum for identifying important contributions and opportunities for research on mass data analysis on microscopic images
- Promote the systematic study of how to apply automatic image analysis and interpretation procedures to that field
- Show case applications of mass data analysis in biology, medicine, and chemistry

Topics

Topics of interest include (but are not limited to):

- Techniques and developments of signal and image producing procedures
- Object matching and object tracking in microscopic and video microscopic images
- 1D, 2D and 3D shape analysis and description
- 1D, 2D and 3D feature extraction of texture, structure and location
- Algorithms for 1D, 2D and 3D signal analysis and interpretation
- Image segmentation algorithms
- Parallelization of image analysis and interpretation algorithms
- Semantic tagging of images from life science applications
- Applications in medicine, biotechnology, chemistry and others
- Applications in crystallography
- Applications in proteomics
- Applications in 2D and 3D cell images analysis
- Image acquisition procedures for mass data analysis

Table of Contents

MDA 2006

New Models for Immune Mechanism Diagnosis

Calin Ciufudean[1,*], Otilia Ciufudean[2,**], and Constantin Filote[1,*]

[1] "Stefan cel Mare" University, University str. 9,
720225, Suceava, Romania
calin@eed.usv.ro, filote@eed.usv.ro
[2] "Areni" Medical Center, Stefan cel Mare str.78, 720229, Suceava, Romania
otilia_ciufudean@yahoo.co

Abstract. In this paper we introduce a discrete event system model of immune systems of mammalians using Markov Decision Processes based on Petri Nets Models. Based on immunological principles we propose an approach in order to study the mechanisms that govern the immune system's functionality. A two-module algorithm that launches a specific action against an anomalous situation is developed. The Petri nets tools are assumed in this approach. Also, Markov Decision Processes (MDPs) with a truncated state space to the problem with infinite state space is considered. We also propose a new algorithm to build a large model (e.g., a macro-model) of immune mechanisms of mammalians. We show that an optimal stationary policy exists and we apply the results of [1] to a dynamic scheduling problem of the immunological response to external stimuli.

Keywords: Petri nets, Markov Decision Processes (MDPs), Immune mechanisms diagnosis.

1 Introduction

Immunity depends on continuous movement of cells through blood, tissue and lymph [2]. Lymphoid cells travel to the secondary lymphoid organs of the spleen, lymph nodes and Peyer's patches to encounter antigens acquired from the environment via blood, lymph or across mucous membranes. Where and by which cells antigens are presented to the trafficking cells has a significant influence on the outcome of the immune response with respect to antibody isotype commitment and future homing preference of memory and effectors lymphoid cells (Fig.1). Lymphocyte traffic patterns, regulated by selective expression of adhesion proteins in peripheral or mucosal lymphatic tissues, permit segregation of immunological memory by causing antigen-primed cells to return to specific anatomic destinations committed to expression of peripheral or mucosal immunity. Among potentially myriad factors, these microenvironments include prevalence of certain cytokines, adhesion to-and co-stimulation by specific cells, and still unknown tissue factors that favor commitment of B cells to specific immunoglobulin types or T cells to peripheral or mucosal immunity.

---

[*] Calin Ciufudean and Constantin Filote are with the Computers and Control Systems Department of the "Stefan cel Mare" University.
[**] Otilia Ciufudean is medical doctor at the "Areni" Suceava Medical Center.

P. Perner and O. Salvetti (Eds.): MDA 2008, LNAI 5108, pp. 1–11, 2008.
© Springer-Verlag Berlin Heidelberg 2008

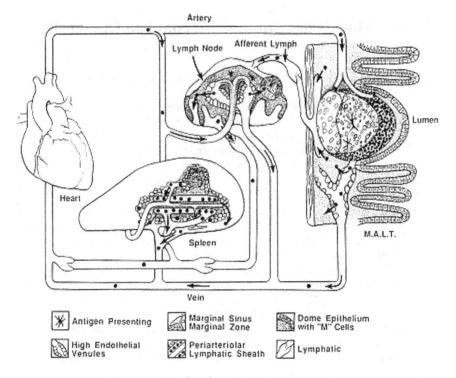

Fig. 1. Mammalians immune system components

Recirculation of a precursor pool of uncommitted lymphocytes from the blood into lymph nodes or mucosal lymphatic tissues and then back to the blood again, integrates immuno-surveillance with organ-selective immune functions across the segregated systems. The magnitude of the cell traffic reflected by the number of cells returned to the blood in efferent lymph is enormous. Enough lymphocytes recirculate from lymph to blood to replace the total blood lymphocyte pool from 10 to 48 times every 24 hours. Random and segregated traffic patterns are essential for efficient operation of the two separate but overlapping immune systems in mammalian species. The feat of coordinating an anatomically dispersed immune system (comprised of mobile, circulating, individual and extremely diverse cells) depends upon cell movement and a system of membrane recognition and activation signals. A mixture of integrins, selectins and chemokine receptors expressed by lymphocytes and endothelial cells are involved in precipitating selective emigration of lymphoid subsets from the blood in tissues where specific counter-receptors are displayed on luminal surfaces of endothelial cells. These recognition events could occur in skin, mucosae or specific secondary lymphatic tissues such as Peyer's patches or peripheral lymph nodes. Receptor ligand interactions allow these cells to find their way around the body, to adhere to endothelium, to migrate and to find where they have to act within the tissues. The cell diversity and variety of information processing mechanisms make the immune system a very complex system. Understanding the way this organ solves its computational, and how it detects and reacts to novel situations and how it unleashes smooth early secondary responses is a rough job. In this paper we present an approach to immune systems by modeling the characteristics processing mechanisms with discrete event systems (DES) formalisms. Our goal is to introduce a new algorithm in order to analyze the organism's fight with viruses and microbes. The proposed algorithm is inspired from the current understanding of the mammal immune system, although, in detail, it does not exactly follow the biological steps. Many of the detailed features of the immune system are dependent on the biological context where it operates and on the type of the cell hardware that it uses. We try to take what is best from the clever evolutionary mechanisms developed by nature, as well as we understand these mechanisms, and to improve their analysis, in order to find new models for treating diseases. For example, the interaction between the T-module and B-module takes the reverse order of what is found in nature, with a clone proliferation phase preceding T-phase. Clone proliferation is an expensive operation, but in software, e.g., in a modeling process, it is a virtual (not very time consuming) operation. The approach presented in this paper has a wide range of applications to many biological, but also to many technical systems. Moreover, based on the optimal policy for the limiting problem built with Markov decision processes (MDPs), we exemplify an optimal stationary policy [3] on a dynamic scheduling response of the immune system to the attack of different pathogen agents.

2 Immune System Mechanisms

Some of the immune system features are [1], [4], [5]:

- Uniqueness: The immune system of each individual is unique.
- Imperfect detection and mutation: By not requiring a precise identification of every pathogen agent, the immune system becomes flexible and increases its detection range. But, if a pathogen agent is detected, a mutation mechanism refines the identification. Identification of pathogen agents is made by partial matching, and this mechanism allows to a small number of the detectors (10^8 to 10^{12}) to recognize non-self patterns on the order of 10^{16}. This is modeled in DES formalism with a small number of detectors, which are at a later stage modified by the dynamics.
- Learning and memory: The immune system can learn the structure of the pathogen agents, and remember those structures. Future responses are much faster and, when made at an early stage of the infection, no adverse effects are felt by the organism. We underline the importance of this feature for modeling the immune system with Petri nets as an important formalism used in the representation of DES.
- Novelty detection: The immune system can detect and react to pathogen agents that the body has never encountered before. This feature will be modeled with controlled Petri nets, which will determine the appearance of bottlenecks in the net, in order to simulate the censoring mechanism for T-cells that occurs in the thymus.
- Distributed detection: The detectors used by the immune system are highly distributed and not subject to centralized control; this feature can be modeled with free choice Petri nets.

3   Modeling Algorithm

Our work is based on the algorithm given in [6]. In this algorithm, the states of the system, both normal and abnormal, are characterized by the values of n variables. The n-dimensional state vector is normalized in such a way that all variables take values in the interval [0,1]. The values of the state vector in normal conditions define the self S of the system. The abnormal states are the non-self of the system. The algorithm adopted by us contains two modules. The T-module discriminates self from non-self. The B-module reacts to all frequently occurring state vector values (self and non-self codes) and reports to the T-module, updating it. T-Module contains a set of detectors which are vectors in non-self space, that is $A = [\,0,1]^n \setminus S$. Each element $\bar{x}$ of A is able to detect anomalies inside a radius r_x around it. When $\left| \bar{y} - \bar{x} \right| < r_x$, y being the current state of the system, an anomaly of type x is reported. In the Petri net model a bottleneck, caused by the fact that an anomaly of type x is not allowed to fire some transition, permits us to emphasize this.

The T-module is initialized by choosing points in A at random with corresponding radius r_x, until a reasonable coverage of the space A is achieved with d detectors. Fig.2.a illustrates this: the small circles are the self patterns. To each point in the self corresponds a code (a set of vector coordinates) and an affinity neighborhood of normal operating conditions inside a radius r_x. This approach corresponds to an initial marking in the Petri net model. The anomaly detectors are shown in the figure as large circles. When a measurement $\bar{y}$ of the system arrives at the T-module, the algorithm verifies whether this code has affinity with one of the detectors or with the self. The affinity of this vector with those defining the self and the other is measured by the Euclidean distance, and correspondingly in the Petri net model is measured with the predicates, assigned to certain transitions, which can or can not validate the firing of the respective transitions.

If the detection algorithm falls in the self domain, no detector is activated.

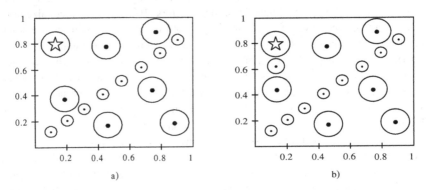

Fig. 2. T-module structure. a) Self patterns (small circles) and anomaly detectors, b) Creation of a new detector and shift to a new detector to increase affinity with an anomaly.

If affinity is found with one of the detectors $\overline{x}$', an anomaly of type $\overline{x}$' is registered. This means that in the Petri net model we'll create new predicates for certain transitions, in order to continue the simulation and to ensure the Petri net vivacity. The B-module generates vector codes corresponding to the most frequently occurring states of the system and sends these codes as alert codes to the T-module. By itself or in interaction with the B-module, the T-module is an adaptive system. As an illustration in Fig. 2.a., and in Fig.2.b, a typical situation is considered: suppose that a non-self code (the star symbol in Fig.2.a) is detected.

First, the detector changes its code to increase the affinity to this type of anomaly, and secondly, the algorithm creates a new detector (supposing that the old one has not enough affinity with the external code) with a resolution defined by the smallest distance to the other detector boundaries as shown in Fig.2.b. In the above way, the T-module modifies the initial set of detectors produced by the censoring mechanism. This means that it changes the number, modifies the space distribution and changes the resolution, creating a specific anomaly detection system. Regarding the Petri net model we may say that we are dealing with an adaptive Petri net (AdPN). B-module improves the A space coverage of the T-module and it has a total population of n_t vectors given by relation (1):

$$n_t = n_l + n_{lc} \qquad (1)$$

Where n_l represents the initial population of vectors $\overline{x_l}$, and n_{lc} represents the population of clone vectors $\overline{x_{lc}}$.

The number of clone vectors changes as the system evolves. In [6] it is allowed that the number of clone vectors is limited to a fraction β of the initial population:

$$n_{lc} = \beta \cdot n_l \qquad (2)$$

The dynamical evolution of the vector population involves mutation and stimulation features that are described next. Mutation takes place every time an external code $\overline{y}$ arrives to the B-module. The mutation process begins by selecting, from the total population, a sample of vectors $\overline{x_m}$. The mutation process operates only in this part of the population and in those codes that are close to the external signal $\overline{y}$. The mutation process depends on the affinity between the vectors $\overline{x_m}$ in the sample and the external code $\overline{y}$. If the code $\overline{y}$ and the vector $\overline{x_m}$ are far away, as in zone A of Fig.3, no affinity is considered to exist and the code $\overline{x_m}$ is not changed. Also, in zone B there is no modification.

For codes $\overline{x_c}$ in zone C, the mutation process occurs in a deterministic way. The external code $\overline{y}$ is assumed to have mass one and the vectors in zone C mass m_l. The new code in population corresponds to the center of mass given in relation (3):

$$\overline{x_c}(t+1) = \frac{m_l \cdot \overline{x_c}(t) + \overline{y}}{1 + m_l} \qquad (3)$$

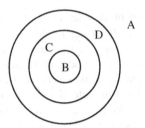

Fig. 3. Zones A-D for the mutation process

For zone D, the mutation assumes a random process. The new position of the population vector $\overline{x_D}$ is given by a random distribution for each point of the line defined by the old position of the vector and the position of the external code (4):

$$\overline{x_D}(t+1) = \overline{x_D}(t) + \eta\left(\overline{y} - \overline{x_D}(t)\right) \tag{4}$$

When the external code appears repeatedly in the same region, the mutation process leads to a population cluster in that region. As we mentioned in section 2, the cluster of population in some regions is modeled with controlled Petri nets, which will determine the appearance of bottlenecks in the net.

Stimulation is a necessary process in the case when new external codes arrive in the B-module and the mutation process destroys the initial uniformity of the vector population. In this situation, if a strange external code appears, its detection may be missed. To ensure that this case is avoided, a stimulation or cloning mechanism has been included in the algorithm to create new vectors in the region where the external code appears. The cloning mechanism is activated when the rate of external codes arriving in a region exceeds a specified threshold. In the Petri net model, this process is modeled adding to the net new location, e.g., building a more complex net. In order to simplify the Petri net model, a pruning algorithm given in [7] is applied. This simplification of the Petri net has a real basis, because there is a death mechanism for the clone vectors in the immune system.

4   Scheduling the Models of the Immune System

The algorithm described in Section 3 does not specify the way of action of the immune system when several extern pathogens occur simultaneously. In order to respond to these pathogens, the immune system needs an action rule similar to the rule of attending to several clients in a queue. The associated Petri net model will be a colored Petri net, where the known pathogens are scheduled in a color code ordered by priorities. The unknown pathogens will be isolated until the known pathogens will be treated by the system. This means that the unknown pathogens will have invalidated entering transitions, because the corresponding predicates are not yet allocated to these transitions.

According to the huge dimensions of Petri net models we try to find out if this scheduling problem has a limit. The answer to this problem is based on the convergence of Markov decision processes (MDP's) with a truncated state space to the problem with infinite state space. In [1] it is shown that an optimal stationary policy exists for this problem, such that the number of randomizations it uses is less or equal to the number of constrains plus one. The following example focuses on this approach. We suppose that different pathogens compete for access to an immune system, which we assume is a shared resource. At the beginning of each time slot priority is given to one of the pathogens according to a pre-specified decision rule, and the service is made in one unit of time (we may consider here the incubation time, which is different for each type of disease). If the service (i.e., the action of the immune system) is successful, the pathogen disappears from the system; otherwise it remains in the queue. The problem is to find a scheduling policy that minimizes a linear combination of the average delays of some types of traffic subject to constrains on average delays of other types. At time t, M_t^i pathogens arrive to queue i, $1 \le i \le N$. Arrival vectors $M_t = \left(M_t^1, ..., M_t^N\right)$ are independent and form a renewal sequence, with finite means λ_i [9], [10]. During a time slot (t, t+1) a pathogen from any class i, $1 \le i \le N$, may be treated, according to some policy, which is a pre-specified dynamic priority assignment. If treated, with probability μ_i it completes its service and leaves the system; otherwise it remains in its queue. A generic element of the state is given by $x = \left(x^1, x^2, ..., x^N\right)$ and it represents an N dimensional vector of different queues' size. Assume that $\sum_{i=1}^{N} \frac{\lambda_i}{\mu_i} < 1$. Consider the linear cost function $c(x,a) = \sum_{i=1}^{N} c_i \cdot x_i$ and

$$d^k(x,a) = \sum_{i=1}^{N} d_i^k \cdot x_i, \text{ for } 1 \le k \le K,$$ where c_i and d_i^k are non-negative constants.

Thus the costs $C(x,u)$ and $D^k(x,u)$ are related to linear combinations of expected average length of the different queues. The constrained control problem is: find $u \in U$ that minimizes $C(x,u)$ s.t. $D^k(x,u) \le V_k$, k = 1, ..., K, where V_k are given constants. Consider the expected average costs. According to Little's law these quantities are proportional to the respective waiting times in the different queues. Let $G = \{g_j\}$ be the set of all strict priority rules. A strict priority rule is a policy for which each type of pathogen is served only if there are no pathogens with higher priority in the system, and if it is the first in his queue. Optimal policies for constrained control problem are obtained by time multiplexing between the different g_j. Define an L dimensional vector parameter $\alpha = (\alpha_1, \alpha_2, ..., \alpha_L)$, where α is a probability measure, and $L = |G|$. Define a "cycle" as the time between two consecutive instants when the system is empty (e.g. the immune system is not busy with external pathogens); during any cycle a g_j is used. A policy α^* is defined as a policy that chooses different policies g_j s.t. the relative average number of cycles during which g_j was used is equal to α_j, where $t \to \infty$. It is shown in [7], [8] that:

$$C_1\left(x, \alpha^*\right) = \sum_{j=1}^{L} \alpha_j \cdot C_1\left(x, g_j\right) \tag{5}$$

For a given d > 0, consider the following linear programming problem:

Find $\alpha \in R^L$ that minimize $\sum_{j=1}^{L} \alpha_j \cdot C_1\left(x, g_j\right)$, subject to $\sum_{j=1}^{L} \alpha_j \cdot D_1^k\left(x, g_j\right) \le V_k - d$,

where k = 1, ..., K, and $\sum_{j=1}^{L} \alpha_j = 1, \alpha \ge 0$.

In [9], [10] it is shown that $\alpha^*(0)$ is an optimal policy for such a constrained control problem. For the Petri net model of the immune system, this means that the initial marking defines the vivacity of the net.

5 Proposal for a Macro-model Algorithm of the Immune System

We have to use the above given approach in order to build macro-models of the mammalian immune system. For that purpose, an iterative modeling process can be described as an ordered sequence of operations $\{W_1, W_2, ..., W_i, ..., W_n\}$. Thus, a model W_i can only be performed after model W_{i-1} and W_{i-2} only after W_{i-3}, and so on. We illustrate our approach via the computation of intruder (microbe, viruses, etc.) cell loss probability in the mechanism described in section 3. We must ensure that our models can simulate the fight of k T-cells against n independent intruder cells sources. We assume that this process is seen as a server which serves a queue with capacity for k cells active (ON) and idle (OFF), represented by 1 and 0, respectively. In the active state, an arrival can occur with probability α. Each of these ON-OFF sources behaves as follows. While an arrival process is in state 1, there is a probability $1-p_{11}$ that it will transit to the idle state at the next time slot and a probability p_{11} that it will remain in state 1, and analogous for the transition probabilities p_{00}, respectively $1-p_{00}$ for the idle state of cells. We assume that a maximum number of c intruder cells can be neutralized in each time slot. The system can be modeled as a discrete time Markov chain [15] with state (x_i, y_i), where x_i is the number of intruder cells in the considered tissue, and y_i is the number of arrival sources in the active state at the time slot number i. Let S denote the state space. Let $T=[t_{m,n;k,l}]$ be the transition matrix for this Markov chain, where $t_{m,n;k,l} = \text{Prob}[x_{i+1} = k, y_{i+1} = 1 \mid x_i = m, y_i = n]$. The dimension of the Markov chain is $(k+1-c)\cdot(n+1)$. The stationary probability distribution can be obtained by solving the equation [16]:

$$\pi \cdot T = \pi \tag{6}$$

The cell loss probability can then be calculated as:

$$P_L = \frac{\displaystyle\sum_{m,n \in S} max(m+n-k) \cdot P[x = m, y = n]}{\displaystyle\sum_{m,n \in S} m \cdot P[x = m, y = n]} \tag{7}$$

Using equation (2), we easily observe that the computational cost is high due to the size of the state space. Therefore we shall use rational interpolants to solve this problem. The major steps required to calculate the rational interpolants for $P_L(k)$ are the following [17]:

1. Suppose

$$log\ P_L(\ k\ ) \cong \beta k \qquad (\ k \rightarrow \infty\ ) \qquad (8)$$

On may calculate the exponential decay rate ß for $P_L(k)$ using the algorithm proposed in [18].

2. Determine the forms of transformation and the form of approximant sequence. Because $P_L(k)$ is exponentially decaying, we develop approximants for the function

$$h(k) = log\ P_L(k) \qquad (9)$$

And will use an $R_{(n+1),n}$ sequence of rational interpolants for h(k), since h(k) is asymptotically linear.

3. Evaluate $P_L(k)$ for small values of k, assuming that the corresponding values of h(k) are known, by solving the Markov chain or using other available analytic methods.

4. Calculate rational interpolants $R_{(n+1),\ n}$ for h(k). This is equivalent to calculating $R_{n,\ n}$ rational interpolant sequence for [$logP_L(k)$ - ßk]. We generate a sequence of rational interpolants, $R_{(n+1),\ n}$ with increasing orders (n = 1, 2, ...) and stop when the successive interpolants are sufficiently close in the range of the considered k.

We notice that a potential macro-model in the course of an iterative process is something that may be the sign of a model to a specialist. A model becomes a component of a macro-model only when submitted to a meditative relation of determination between the subject of the model and the specialist. We propose an algorithm to perform macro-models. It presupposes the notion of environment (e.g., subject of the model) and the specialist [12-14]. The synthetic environment represents the reality that is forced upon the specialist's expertise. The environment is infinitely complex (from the view point of specialists). Specialists, which are immersed in the environment, are able to perceive and act on the environment. The proposed algorithm is the following one:

a) Choose a collection of models (potential components of the macro-model): M = {M_i}, i=1,..., n.
b) Choose a model M_k, k=1, ..., n from this collection.
c) Propose a potential synthetic environment and specialist, so that there is a relation which is proved to be functional by the simulation of the syntactic structure of M_k, k=1, ..., k-1. Then we say that M_k determines the macro-model relatively to that specialist and to the previous model M_{k-1}.

In order to implement this algorithm, one must first define some sort of cognitive architecture for the agent, in which sensors and effectors are specified. The goal of this proposal is to state very basic steps to perform simulated macro-models of the immune systems of mammalians.

6 Conclusion

In time, nature's evolutionary processes created an efficient weapon to fight with all kinds of hostile environments. Modeling these natural mechanisms seems to be a sensible approach. In this paper we proposed a possible tool for this approach: Petri nets. However, some of the features of the biological processes are domain specific and depend on the cell hardware that is used. Therefore, understanding and modeling these processes are hard tasks.

The immune system, with its cell diversity and variety of information processing mechanisms, is a very complex system. The high complexity of the immune system implicates very large Petri net models. In order to minimize the dimensions of models, we introduce the notion of adaptive Petri nets. Therefore we have shown that there is a limit in the schedule problem of different pathogens, which compete to access a limited service capacity of the immune system. For this we considered the Markov Decision Processes with a truncated state space to the problem with infinite space. Future work will refine the above presented approach by considering differential adaptive Petri nets for modeling the mechanisms which govern the immune system; this is motivated by the necessity to model certain nonspecific mechanisms like cell apoptosis, etc., which was not discussed here.

References

1. Altman, E.: Asymptotic properties of constrained Markov decision processes. SIAM J. Control and Optimization 29(4), 786–909 (1992)
2. Gretz, J., Anderson, A., Shaw, S.: Cords, channels, corridors and conduits: critical architectural elements facilitating cell interactions in the lymph node cortex. Annual Revue Immunology 15, 11–24 (1997)
3. Doherty, P.C., Christensen, J.P.: Accessing complexity: The dynamics of virus specific T-cell responses. Annual Revue Immunology 18, 561–592 (2000)
4. Banchereau, J., Biere, F., Caux, C., Davoust, J., Lebeque, S., Liu, Y., Pulendran, B., Palucka, K.: Immunobiology of dendritic cells. Annual Revue of Immunology 18, 767–811 (2000)
5. Timmis, J., Neal, M., Hunt, J.: An artificial immune system for data analysis. Biosystems 55, 143–150 (2000)
6. Costa Branco, P.J., Dente, J.A., Vilela Mendes, R.: Using immunology principles for fault detection. IEEE Trans. on Ind. Electr. 50(2), 362–372 (2003)
7. Recalde, L., Teruel, E., Silva, M.: Modeling and analysis of sequential processes that cooperate through buffers. IEEE Trans. on Rob. and Autom. 14(2), 267–277 (1998)
8. Ciufudean, C.s.a.: Intelligent Control of Artificial Social Systems. In: Proc. 8th International Conf. on Development and Application Systems, Suceava, May 25-27, pp. 115–119 (2005)
9. Ciufudean, C., Filote, C.: Performance Evaluation of Distributed Systems. In: International Conference on Control and Automation, ICCA 2005, Budapest, Hungary, June 26-29, pp. 21–25 (2005), IEEE Catalog Number: 05EX1076C, ISBN0-7803-9138-1
10. Ciufudean, C., Filote, C., Amarandei, D.: Measuring the Performance of Distributed Systems with Discrete Event Formalisms. In: Proc. of The 2nd Seminar for Advanced Industrial Control Applications, SAICA 2007, Madrid, Spain, November 5-6, pp. 263–267 (2007), ISBN 978-84-362-5519-5

11. Ciufudean, C., Graur, A., Filote, C., Turcu, C., Popa, V.: Diagnosis of Complex Systems Using Ant Colony Decision Petri Nets. In: The First International Conference on Availability, Reliability and Security, Vienna University of Technology, pp. 35–39 (2006), IEEE Catalog Number: 05EX1076C, ISBN0-7803-9138-1
12. Houser, N.: Introduction: Peirce as a logician. In: Houser, N., Roberts, D., Evra, J. (eds.) Studies in the logic of Charles Sanders Peirce, pp. 1–22. Indiana University Press (1997)
13. MacLennan, B.J.: The emergence of communication through synthetic evolution. In: Patel, M., Honavar, V., Balakrishnan, K. (eds.) Advances in the Evolutionary Synthesis of Intelligent Agents, pp. 65–90. MIT Press, Cambridge (2001)
14. Perfors, A.: Simulated evolution of language: A review of the field. Journal of Artificial Societies and Social Simulation 5(2), 35–41 (2002)
15. Nananukul, S., Gong, W.B.: The mean waitinh time of GI/G/1 queue in light traffic via random thinning. J. Appl. Probability 32, 256–266 (1995)
16. Assmussen, S.: Applied Probability and Queues. John Wiley & Sons, Chichester (1987)
17. Nananukul, S., Gong, W.B.: Rational interpolation for stochastic DES's: Convergence issues. IEEE Trans. On Autom. Contr. 44(5), 1070–1073 (1999)
18. Gong, W.B., Nananukul, S.: Rational approximants for some performance analysis problems. IEEE Trans. Comput. 44, 1394–1404 (1995)

User Assisted Substructure Extraction in Molecular Data Mining

Burcu Yılmaz[1,2], Mehmet Göktürk[1], and Natalie Shvets[1]

[1] Gebze Institute of Technology, Dept. of Computer Science, Turkey
[2] Istanbul Kultur University, Dept. of Computer Engineering, Turkey
b.yilmaz@iku.edu.tr, {gokturk,natali}@gyte.edu.tr

Abstract. In molecular fragments mining, scientists use both manual techniques and pure computer based methods. In this paper, we propose a novel molecular fragment mining approach that incorporates interactive user assistance to speed up and increase the success rates in traditional fragment mining processes. The proposed approach visualizes 3D molecular data in 2D form that can be easily interpreted by a human expert who evaluates and filters the 2D molecular images manually. The proposed approach differs from others in literature as it does not search substructures including specific atoms like graph mining methods do. Instead, user assisted approach highlights significant substructures with specific properties and topologies graphically. Initial experiments indicate that by the use of user assisted approach, active and inactive fragments of compounds are quickly determined for drug design with high success rates.

Keywords: Molecular mining, interactive visual data mining.

1 Introduction

Design of new medical drugs is a time consuming and expensive process. The success depends on the molecular representations and methods used for the common active fragments selection. Researchers study the relationships between structure and desired activity of the compounds against a specific disease. The studies are known as Structure Activity Relationships (SAR) investigation.

All approaches used in drug research have some common steps. First, based on experimental evidence of different degrees of activity against a certain disease, a large set of drug candidates is determined and labeled. Then, those candidates are analyzed manually by designers or automatically by computer programs. Hundreds of compounds need to be analyzed to find rules, correlations or substructures for desired pharmacological activity. Many approaches are developed for common problems such as finding common substructures or classification of molecules according to their chemical activity. These approaches are grouped into two categories in the literature, Quantitative SAR and Qualitative-SAR.

Some of the researchers in drug design base their approaches on the principle that the whole molecular structure (not substructures) of an active molecule is responsible for the activity. These approaches are known as Quantitative SAR, which derives a correlation between the molecular descriptors and values of activity. Molecular

P. Perner and O. Salvetti (Eds.): MDA 2008, LNAI 5108, pp. 12–26, 2008.

structure descriptions are prepared for every molecule as vectors. Then different machine learning techniques are applied to the prepared dataset as described in the literature. Linear discriminant analysis, multilayer perceptrons, support vector machines, k-nearest neighbors, linear regression, classification tree are the most studied methods on Quantitative SAR for the molecules classification [1].

Remaining part of the research community believe that several substructures are responsible from the desired activity and they propose methods known as Qualitative-SAR which are implemented through finding some common frequent substructures (active fragments) in molecular structure. An active fragment (pharmacophore) is the most frequent similar structural feature in the structures of active molecules (or drugs), which is responsible for the molecule's biological activities. There are many approaches related to Qualitative-SAR in the literature. In Inductive Logic Programming (ILP) approach, molecules are represented using first-order logic [2]. ILP derives logic based rules to identify combinations of the features belonging to active fragments. Some of the researchers model and visualize compounds as 3D graph representations using graph theory and studies on frequent graph mining approaches [3] to find frequency of common fragments (pharmacophores) in compounds with same activity for a specific disease. In this representation atoms and bonds are represented as some specific descriptors. The atoms of molecules are represented by vertices of graphs and the bonds are represented by edges. Some Qualitative-SAR methods use graph mining methods to find common frequent substructures that exist in maximum probability of active molecules and minimum probability of inactive molecules. They also try to avoid the complexity of two main problems in graph theory. The decision problem of whether two graphs have identical topological structure has an unknown complexity (Graph isomorphism) [4]. The second problem that is deciding whether one graph is a subgraph of another (subgraphs isomorphism) is known to be NP-complete [4]. One of the well known approaches which uses greedy search to avoid high complexity of graph isomorphism is *SUBDUE* [5]. However, only an incomplete set of frequent substructures can be found due to greedy search. Some methods like *FSD, gSpan, AGM* create candidate subgraphs from each molecule by adding one edge to each candidate each time [6-9]. Then isomorphisms of each candidate at different molecules are searched in dataset. To prune the search space, some of the candidate subgraphs are eliminated using some limitations, heuristics or support values. Instead of using some limitations for searching candidate subgraphs to avoid curse of graph and subgraph isomorphism, clustering methods can be another feasible alternative. On the other hand, direct use of clustering methods may cause wrong clusters because of the scattered data of flexible molecules. To avoid this deficiency, we propose to give pre-information from the experts about fragments which will increase the success, and prune the search space.

In this paper, the main focus is Qualitative-SAR method that uses pharmacophores as a map to guide for molecular design. The proposed novel molecular fragment mining approach uses suggestions of experts instead of fully automated blind computer calculations. With the proposed approach, chemists use 2D molecular information visualization to filter data. 2D Molecular visualization techniques are used to create molecular images and these images are used to filter uninformative regions of the molecules. The novel representation displays common properties of active and inactive molecules as "filtered-out" parts. With a simple *data flow* technique, the data is filtered through several molecular visualization techniques with variable parameters until the data is filtered satisfactorily. The filtering method is basically the first stage of feature extraction method where residues give more exact information about active fragments. Second stage of feature extraction determines the final active fragments. This interactive user assistance in the proposed approach speeds up the process without any decrease in success rate.

2 Methodology

In this section, we explain our methodology for molecular fragment mining in detail. For this purpose, we introduce a semi-automatic pipelining structure in the next subsection.

2.1 Semi-automatic Data Pipelining Structure & Data Filtering

Previous researches mostly use data mining techniques to find active fragments [3]. Parameters of these techniques are automatically determined using adaptive approaches.

The success of these systems depends on the accuracy of the blindly estimated parameters [10]. Since finding active fragments is among the most important stages of medical drug design, the parameters like the cluster centers and size of activity clusters should be selected with extreme care and accuracy instead of totally depending on automatic algorithms to determine the parameters.

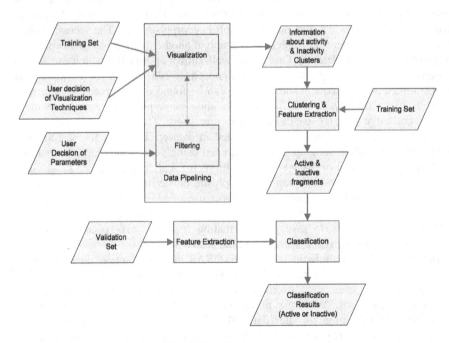

Fig. 1. Data flow diagram

In the proposed methodology, the most important step of molecular fragment mining is filtering. It is also vital stage of feature extraction. In our study, we have used a *data pipelining* structure to enable an expert to select the data that needs to be filtered. Figure 1 shows the data flow diagram of this data pipelining structure. Raw molecular data is filtered using different phases iteratively. During the process, the expert provides critical information to the system. For example, the expert determines the degree of filtering by providing suitable parameters to the system by examining visual representations presented. After getting suitable parameters from the expert, the system extracts features from the molecules' data. Data pipelining phase consists of several filtering stages. After filtering, final activity clusters are identified clearly for the active fragment extraction from the molecular data.

2.2 Visualization of Molecule Properties

Graphs are one of the most frequently used techniques in molecule visualization [3]. However, this visualization technique can indicate only information about topology of molecules. Although edges and nodes are used to represent some properties of molecules, it is inefficient to compare different properties of 3D molecules with graphs only by visualization.

In this paper, a transformation from 3D graphs to 2D molecular information visualization is implemented. To make this representation more understandable to the experts, images that display information about molecules are created. These images contain topological information of molecules as well as additionally requested properties. To represent molecular graphs, *Electron-Topological Matrices of Conjugency* (ETMC) are used [11]. In this method, a molecule with n atoms is represented with an upper triangular ETMC matrix as shown below:

$$ETMC\ Matrix = \begin{bmatrix} a_{1,1} & a_{1,2} & \cdots & a_{1,n-1} & a_{1,n} \\ & a_{2,2} & \cdots & a_{2,n-1} & a_{2,n} \\ & & \cdots & \cdots & \cdots \\ & & & a_{n-1,n-1} & a_{n-1,n} \\ & & & & a_{n,n} \end{bmatrix} \qquad (1)$$

In our study, diagonal elements a_{ij} $(i=j)$ of an ETMC matrix contain information about electronic properties of atoms and non-diagonal elements a_{ij} $(i \neq j)$ include information about chemical properties of bonds between the corresponding atoms in ETMC matrix representation. If there is no bond, the distance between two atoms is used instead. The elements of ETMC matrix can vary, so that different properties of molecules are examined for effective properties on activity.

For every integer values of i, j $(1 \leq i, j \leq n)$, vectors $[a_{ij}, min(a_{ii}, a_{jj}), max(a_{ii}, a_{jj})]$ are obtained from ETMC matrices, so that a molecule is split into pieces (bonds), each bond is represented with a vector including information about bonds (a_{ij}) and two atoms at each ends (a_{ii} and a_{jj}). These pieces are plotted with point pairs, (a_{ij}, $min(a_{ii}, a_{jj})$) and (a_{ij}, $max(a_{ii}, a_{jj})$) using a transformation from 3D to 2D coordinate system as shown in Figure 2. Thus, points are derived from an ETCM matrix, and then can be plotted into 2D cartesian system. In the following sections, we propose two different approaches for the molecule visualization.

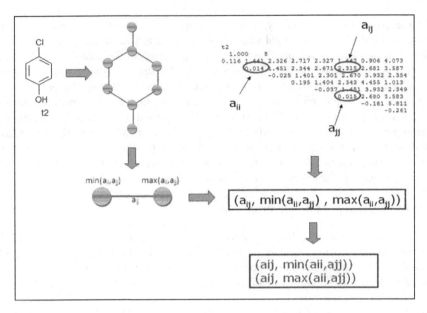

Fig. 2. Transformation of data to 2D space

2.2.1 Gray Scale Shading for the Molecule Visualization

We propose a visualization technique that displays desired properties of molecules for the experts using 2D images. Therefore, we enhance the understandability of the molecular information by experts and enable the experts to compare different molecules easily in terms of their desired properties. One of the problem of comparing molecules is that molecules posses different degrees of flexibility. Because of this flexibility, topological structures of same molecules can vary under different conditions. The ETMC matrix values therefore cannot be compared in a straightforward manner. To overcome this difficulty, we decrease the resolution of the data as mentioned below. A 2D image is generated from the ETMC matrix of a molecule by dividing X-Y coordinate system into regular intervals in order to compose a 2D mesh. For this purpose, two parameters are used; Δ_x and Δ_y. Using the parameter Δ_x, X dimension of the coordinate system is divided into the intervals so that the width of each interval is Δ_x. Similarly, this is repeated to the Y axis. The crossing intervals on the coordinate system compose a 2D mesh. The resulting mesh is used to create a visualization of the molecule. Each point drawn from the ETMC matrix is plotted into the 2D mesh. Then, the rectangle region of the mesh where the point falls into is shaded.

Two different techniques are used in shading rectangle regions. First technique paints the rectangle region into black, if the point falls into it (see Figure 3). Figure 4 shows an example for the creation of a 2D image from an ETMC using the first technique.

In the second technique, the color of the rectangle and the colors of its neighbor rectangles represent the position of the point in the corresponding rectangle region (see Figure 5).

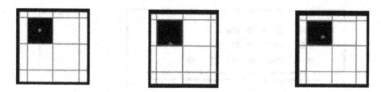

Fig. 3. Image formation using the first technique

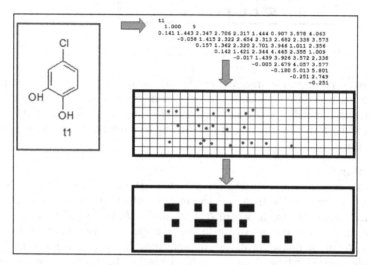

Fig. 4. Resulting 2D image after processing an ETMC matrix according to the first technique

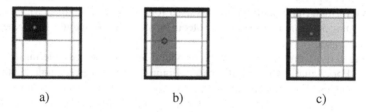

a)   b)   c)

Fig. 5. Image formation using the second technique

Hence, unlike the first technique, the second approach enables experts to preserve more information about the position of the points in the 2D coordinate system. This technique is based on *antialiasing* method that is used to adjust the distortion of information (aliasing) due to low-frequency sampling [12]. As the first technique decrease the resolution of the data, the second technique avoids the data loss incurred using an *antialiasing*-based methodology. For example at the second technique, if the point is close to a border, that rectangle and neighbor rectangle is painted to gray according to the closeness of the point to the border (Figure 5-b). If the point is close to two borders, the color of the rectangles and three neighbor rectangles are calculated according to the closeness of the point to the borders of the rectangles (Figure 5-c). If the point is at the center of the rectangle, that rectangle is painted to black(Figure 5-a).

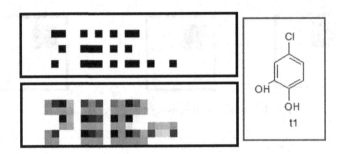

Fig. 6. Molecule images of molecule t1 using two shading methods

For every point pair extracted from ETMC matrix of a molecule, intervals are shaded using two different shading techniques mentioned above. The comparison of two different methods for the same molecule is shown in Figure 6. 2D molecular images can be formed using one of the shading techniques, depending on the choice of the expert.

Activity Map
An activity map is created using the 2D mesh. However, this time, shades of the rectangles in the activity map are determined differently. A rectangle $R_{I,J}$ can be considered in the 2D mesh, where $0<I<N$ and $0<J<M$. In order to determine the shade of each $R_{I,J}$ in the activity map, we average the shade of the corresponding rectangle $R_{I,J}$ in the images of the active molecules. In order to average shades, 8-bits gray scale is used where 0 corresponds to black. The resulting average value is converted to a gray scale. Hence, an activity map gives information about the parts of molecules that appear frequently in the structure of active molecules in a better perceivable format. Same methodology is used for creation of the inactivity maps, but this time, images from the inactive molecules, instead of images from the active molecules are used.

Resulting activity and inactivity maps provide important perceivable information to the experts. For example, the activity map in Figure 7 shows the rectangles with darker and lighter shades. The dark fragments are the structures that are common to the most of the active molecules whereas the lighter fragments represent the structures that are not repeating in the active molecules, so those fragments may not be significant for the activity. Hence, we filter the training data falling to the lighter fragments in the activity map using a threshold value determined by the expert. In summary, first activity maps are constructed using all active molecules in the training data, then algorithm determines the insignificant points of the ETMC matrix of each molecule in the training data using the decisions of expert and lastly filters those points to eliminate redundant and noisy information. Alternatively, data falling into some fragments on the image can be removed directly by the expert. The same procedure can be applied to the inactivity map. Dark fragments of the activity map and inactivity map may also be filtered if those fragments exist in both images. This is intuitive, because those common fragments in the activity map and inactivity map do not give any significant information about active fragments. Selection of Δ_x and Δ_y parameters don't effect the

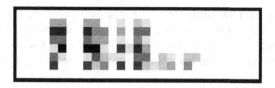

Fig. 7. The activity image of 10 active molecules

extracted active fragments too much, because the filtered data gives only preliminary information about activity and inactivity clusters. Nevertheless different values of the Δ_x and Δ_y parameters should be tried for best filtering.

2.2.2 HSV Scale Coloring for the Molecule Visualization

Gray scale shading enables expert to create activity and inactivity maps. Some fragments of these maps should be removed in order to get active and inactive fragments. As stated before, expert assistance is used. For this purpose, expert provides a threshold value. The expert uses only a gray scale image (map) to decide on a threshold value to provide this assistance. Perception and expertise of the human expert is therefore crucial. Different colors may be easily perceived as the same by the human even if they are significantly different colors. Perceptual capacity of human in discriminating gray scale values is less than the capacity in discriminating color in hue values [12].

In this section, we propose a second coloring method for the molecule visualization. In this method, we use the same transformation from 3D to 2D as mentioned in Section 2.2.1. That is, from the ETMC matrix of a molecule, we derive 2D points and we project those points on a 2D mesh that is determined by the parameters Δ_x and Δ_y. An activity map using the 2D mesh is computed using a rectangle $R_{I,J}$ on 2D image, where $0<I<N$ and $0<J<M$. In order to determine the color of the each rectangle $R_{I,J}$ on the activity map, first the color of its center is determined. For this purpose, a simple voting algorithm is used. In this algorithm, each active molecule votes for $R_{I,J}$ if one of the points derived from the ETMC matrix of the molecule falls on $R_{I,J}$. Similarly, each inactive molecule votes for the rectangle $R_{I,J}$ using the same criteria. Percentage of the active and inactive molecules that vote for the rectangle $R_{I,J}$ is denoted as $P^A_{I,J}$ and $P^{IA}_{I,J}$. If a rectangle is voted by most of the active molecules but a small percentage of the inactive molecules, then this rectangle may contain features that are significant for being an active molecule. In order to highlight those rectangles, we compute the difference of the computed percentages as shown below.

$$D_{I,J} = P^A_{I,J} - P^{IA}_{I,J} \qquad (2)$$

The computed difference determines the color of the rectangle's center. For this purpose, we use a color map that maps $D_{I,J}$ values to corresponding colors on Hue Saturation Value (HSV) scale where colors are based on an intuitive color model with spectral colors [12]. Color distribution on this map is chosen so that colors mapped to different hues can be differentiated easily by human experts. After determining the colors of the rectangle centers this way, those colors are used to determine the color distribution on the map by using *2D color hue interpolation*. For example, color distribution within the rectangle $R_{I,J}$ is computed using the interpolation of 9 colors; its center' color and centers' color of its 8-neighbours. 8-bit hue values *0-255* where active parts correspond to higher hue values towards the red edge of the spectrum.

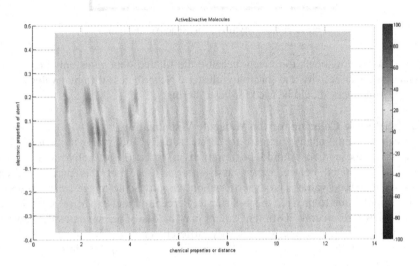

Fig. 8. Activity-Inactivity Map of HSV Scale Coloring

In the computed activity-inactivity map in Figure 8, the higher degree of red color shows the higher percentage of activity so that only points falling into these parts give active fragments. A similar procedure is used for inactive molecules. The higher degree of blue color shows the higher percentage of inactivity at those parts. Green parts that have zero percentage of activity and inactivity don't have a role in activity or inactivity.

2.3 Extracting Fragments & Clustering

After filtering the training data using active and inactive maps (mentioned in 3.2.1 and 3.2.2), redundant and noisy data are removed from the dataset. We determine active fragments of the molecules as follows. First, molecular graphs are transformed into a 3D space where each point corresponds to bonds splitted from molecules including properties of the bond and two atoms at each two ends, so that a molecule with n atoms and m bonds is represented with m points: $(a_{ij}, min(a_{ii}, a_{jj}), max(a_{ii}, a_{jj}))$. This way, all the molecules are transformed into the same 3D atom-bond-atom space. So that, intensified point groups forms activity or inactivity cluster candidates, each denote active or inactive fragment candidates. Then, *average-link clustering* method is used to find candidates of clusters [13, 14]. In this clustering method, initially each point in the space is regarded as an individual cluster. Then, clusters are merged iteratively according the distances between the cluster centers. That is, two clusters are merged if their distance is smaller than a predefined threshold. Note that each cluster is composed of one or a set of bonds. After determining clusters, for each cluster, we compute the percentage of active and inactive molecules that

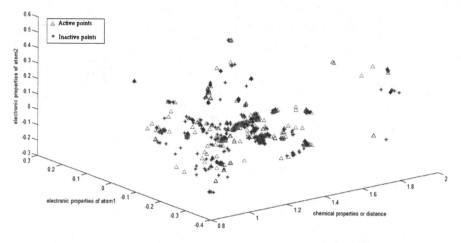

Fig. 9. Unfiltered active molecules plotted on atom-bond-atom coordinate system

contain the molecular pieces in the cluster. The clusters including higher percentages of active molecule pieces and small percentage of inactive molecule pieces are regarded as activity clusters. Bonds falling into active clusters are expected to be active fragments.

In order to show advantage of the proposed filtering method, we show the points extracted from raw training-data in Figure 9 and the filtered training-data in Figure 10 in the 3D atom-bond-atom space. Although Figure 9 gives a messy view of the training data, Figure 10 gives a cleaner view of the training data, because noise and uninformative points are removed from the training data after filtering. In Figure 10, we also show the resulting clusters on the filtered training data, and an example molecule's two pieces falling into two activity clusters.

3 Results and Discussion

In order to demonstrate our approach better, we design realistic experiments with real-life data. Cost of labeling molecules as active and inactive is usually high. It requires extensive analysis in laboratory conditions. Naturally, most of the analyzed molecules are found to be inactive for a specific disease. Therefore, in real-life settings, datasets are small and most of the molecules in those datasets are inactive. Hence, molecular data mining is very hard in real-life settings. Especially mining active molecules are relatively harder than mining inactive molecules, because we usually have relatively less information about the active molecules.

In our experiments, we use anti-tuberculosis dataset [15]. This dataset is composed of 33 molecules (13 active and 20 inactive molecules). All the experiments of the proposed method are tested on 1.6 Ghz Intel Pentium Core II Duo, 1 Gb Ram running Windows Vista operating system. To make the approach repeatable, we used Matlab 7.1 Prtools Toolbox for classification methods. We demonstrate the performance of our approach in two steps. First, we show a case study to demonstrate how an expert can use our approach for creating activity & inactivity maps and deriving active and inactive fragments interactively. Second, we use the found fragments as features and

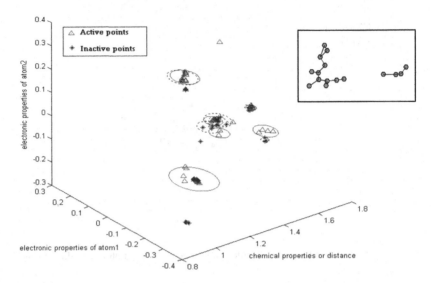

Fig. 10. Activity and inactivity clusters extracted from filtered data and active fragments extracted from the clusters

evaluate the performance of well-known classification methods in classifying molecules as active or inactive. Intuitively, if our approach is successful in determining active and inactive fragments that account for the activity and inactivity of the molecules, then the classification methods using those fragments as features are expected to demonstrate a good performance.

3.1 Case Study

An expert is first asked for the parameters Δ_x and Δ_y. Those parameters are selected as $\Delta_x=0.12$ and $\Delta_y=0.07$ by the expert. Using those parameters, activity and inactivity maps are created as in Figure 11 and Figure 12, respectively. Using those activity and inactivity maps, the training data is filtered using a threshold value: % 40. This threshold value is decided by the expert in order to keep more information unfiltered. Although it may seem that using a 2D activity maps rather than a 3D representation may cause loss of data. Important to note that, it is only used in finding approximate locations of clusters where unfiltered data and preinformation about clusters were fully preserved. Lastly, using unfiltered data and pre-information about the clusters, the system finds final active fragments. To visualize 3D topology of active fragments, an active template molecule which is selected by experts is used. Graph based representation of extracted active fragments on the template molecule are shown at figure 13.

During these experiments, finding only clusters and fragments (not including active map formation) using $\Delta_x=0.12$ and $\Delta_y=0.07$ takes 9.5 seconds, and total processing time from beginning to finding active fragments is around to 11 minutes and 12 seconds.

Fig. 11. Activity map

Fig. 12. Inactivity map

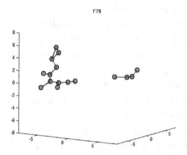

Fig. 13. Active fragments

We also measure processing time for image creation and filtering (Feature extraction) for different values of the parameters Δ_x and Δ_y. Table 1 show time consumption for *Gray Scale Shading (GSS)* and *HSV Scale Coloring (HSVSC)*, respectively. The table shows that the parameters Δ_x and Δ_y do not significantly affect processing time for *GSS*'s image creation, while they dramatically affect processing time required for the creation of activity & inactivity maps for both *GSS and HSVSC*.

These tables also show that *HSVSC* requires much more time than *GSS*, because of the interpolations. Therefore, while *HSVSC* improves the perception of the experts by producing colorful maps and images, *GSS* has a much better response time. For different experts, those methods may have different advantages and disadvantages. Hence, each expert may decide on one of those methods depending on its trade-offs (e.g., perceptional improvement vs. response time). Additionally, we may note that both of these methods are fast enough to enable extraction of active and inactive fragments within minutes, during our experiments.

Table 1. Process times for Gray Scale Shading and HSV Scale Coloring

		Gray Scale Shading (GSS)		HSV Scale Coloring (HSVSC)
Δ_x	Δ_y	Image creation (sec.)	Act. & inact. map creation (sec.)	Act. & inact. map creation (sec.)
0.12	0.07	44.16	16	129
0.4	0.15	43.5	9	69

3.2 Classification of Molecules Using Active/Inactive Fragments as Features

Searching a molecular substructure rapidly in a molecular database is an important research problem in drug design. In the literature, graph mining techniques are mostly used to search molecular databases for the new molecules that are likely to be active for a specific disease [16]. Using classical graph searching methods, it is difficult to find molecules with specific frequent substructures because of the time complexity of subgraph isomorphism. Instead of using graph searching methods, classification methods from the machine learning literature can also be used for the estimation of active and inactive molecules in a molecular database. Classification methods require each instance in a training set to have the same dimensions. Therefore, we first derive a set of active and inactive fragments. Then, we use these fragments as features. This way, we represent each molecule using the same dimensions as follows. Let we have n_1 active and n_2 inactive fragments that we derive using the method in section 2.3. For each molecule, we prepare an array of n_1+n_2 dimensions, where each dimension represents one fragment. If the molecule has an active or inactive fragment, the corresponding dimension of the vector is set to the minimum distance from the points of molecules to the corresponding cluster's center; otherwise it is set to 0. This way, we represent each molecule with the same dimension vectors. Using this methodology, we create a training set from the labeled examples. Then, we input this training set to different classifiers.

In the QSAR literature, different classifiers are used. Most popular classifiers are *Decision trees (DT), Linear Discriminant Analyses (LDA), K-Nearest Neighbor (K-NN), Support Vector Machines (SVM)*. In our work, we also use these popular classifiers. We train these classifiers using the training set that is prepared using the derived active and inactive fragments. Then, using *one-leave-out* cross validation method, we measure the classification performance of the classifiers.

In order to measure the performance of our approach better, we also train these classifiers with the original unfiltered data that do not only contain active/inactive fragments but also other fragments that are filtered out while determining these active/inactive fragments. Unfiltered data contain more information about the molecules, so classifiers using the unfiltered data may have a better performance with respect to their performance using only a subset of this data (i.e., only active and inactive fragments). Our main aim in this paper is correctly determine active/inactive fragments that are responsible for the activity and inactivity of the molecules. If we determine these fragments correctly, the classifiers using only those fragments as features should also demonstrate a good performance in classifying molecules.

In Table 2, we tabulate our results for $\Delta_x = 0.12$ and $\Delta_y = 0.07$ using the original (unfiltered) data and the filtered data, where only active and inactive fragments are used as features. Our results show that classifiers achieve the same performance when they use only active/inactive fragments as features (filtered data) and when they use the whole molecular data (unfiltered data). In our experiments, the best performance belongs to *SVM* and *DT* classifier, which always correctly classifies active and inactive molecules (success is *100%*). Performance of other classifiers are also very good, almost all of the classifiers can correctly classify molecules more than *95%* of the cases. Those results imply that our approach can correctly determine the active and inactive fragments and those fragments can successfully be used as features in

Table 2. Performance of the classification methods on the antituberculosis dataset

	Unfiltered Data		Filtered Data	
	Active Molecules	Inactive Molecules	Active Molecules	Inactive Molecules
Classification Methods	Success (%)	Success (%)	Success (%)	Success (%)
Decision Tree	100	100	100	100
Linear Discriminant Analysis	0	100	92	95
1-Nearest Neighbor	100	100	100	95
Support Vector Machines	100	100	100	100
Average Processing Time	39 min 19 sec.		14 min 29 sec	

classification. Moreover, when the classifiers use only the active/inactive fragments extracted from filtered data rather than fragments extracted from original (unfiltered) data, it is observed that overall classification process is 2.7 times faster.

4 Conclusions

Machine learning methods use blind calculations to estimate necessary parameters. However, incorrect parameter estimations can adversely affect the results. Every molecular dataset has various structures, so the models fitted on these datasets can differ. Most of the current methods in the literature depend on the structure and size of molecules. They try to avoid the complexity of graph and subgraph isomorphism with some limitations in search space. However, in this paper, we present a novel approach to find substructures with common properties to avoid these deficiencies.

We have evaluated the performance of our approach using experiments. Our experiments show that our approach can correctly determine active and inactive fragments of molecules that account for the activity and inactivity of those molecules. We also show that using our approach, classification methods can achieve good performances while determining active and inactive molecules.

In this work, we do not directly compare our approach with the other methods in the Qualitative-SAR literature, such as sub-graph searching based approaches. As a future work, we are planning to compare our approach experimentally with the other approaches from the literature. A comprehensive study on unbalanced datasets can further enhance the future research planned [17].

References

1. An, A., Wang, Y.: Comparison of Classification Methods for Screening Potential Compounds. IEEE, Los Alamitos (2001)
2. Stenberg, M.J.E., Muggleton, S.H.: Structure Activity Relationships (SAR) and Pharmacophore Discovery Using Inductive Logic Programming (ILP). QSAR Comb. Sci. 22 (2003)
3. Fischer, I., Meinl, T.: Graph Based Molecular Data Mining – An Overview. IEEE international Conference on Systems, Man and Cybernetics (2004)

4. Yang, G.: The Complexity of Mining Maximal Frequent Itemsets and Maximal Frequent Patterns. In: ACM SIGKDD international conference on Knowledge discovery and data mining (2004)
5. Cook, J., Holder, L.: Substructure discovery using minimum description length and background knowledge. J. Artificial Intel. Research 1, 231–255 (1994)
6. Inokuchi, A., Washio, T., Motoda, H.: Complete mining of frequent patterns from graphs:Mining graph data. Machine Learning 50, 321–354 (2003)
7. Kuramochi, M., Karypis, G.: Frequent subgraph discovery. In: ICDM 2001: 1st IEEE Conf. Data Mining, pp. 313–320 (2001)
8. Yan, X., Han, J.: gspan: Graph-based substructure pattern mining. In: ICDM 2002: 2nd IEEE Conf. Data Mining, pp. 721–724 (2002)
9. Deshpande, M., Kuramochi, M., Wale, N., Karypis, G.: Frequent Substructure-Based Approaches for Classifying Chemical Compounds. IEEE Transactions on Knowledge and Data Engineering 17(8) (2005)
10. van der Heijden, F., Duin, R.P.W., de Ridder, D., Tax, D.M.J.: Classification, Parameter Estimation and State Estimation. Wiley, Chichester (2004)
11. Shvets, N., Dimoglo, A.S.: The Electron-topological method (ETM): Its further development and use in the problems of SAR study. In: Molecular modeling and prediction of bioactivity, pp. 418–429. Kluwer Academic/Plenum Publishers, New York (1999)
12. Baker, H.: Computer Graphics, 3rd edn. Prentice-Hall, Englewood Cliffs (2003)
13. Alpaydın, E.: Introduction to Machine Learning, 1st edn. MIT Press, Cambridge (2001)
14. Xu, R., Wunsch, D.: Survey of Clustering Algorithms. IEEE Transactions on Neural Networks 16(3) (May 2005)
15. Nayyar, A., Monga, V., Malde, A., Coutinho, E., Jain, R.: Synthesis, anti-tuberculosis activity and 3D-QSAR study of 4-(adamantan-1-yl)-2-substituted quinolines. Bioorganic&Medicinal Chemistry 15, 626–640 (2007)
16. Nijssen, S., Kok, J.N.: Frequent Graph Mining and its Application to Molecular Databases. In: IEEE International SMC (2004)
17. Molinara, M., Riamato, M.T., Tortorello, F.: Facing imbalanced classes through aggregation of classifiers. In: IEEE ICIAP 2007, pp. 43–48 (2007)

Fully Automatic Heart Beat Rate Determination in Digital Video Recordings of Rat Embryos

M. Khalid Khan[1], Mats F. Nilsson[2], Bengt R. Danielsson[2], and Ewert Bengtsson[1]

[1] Centre for Image Analysis, Uppsala University, Sweden
{khalid,ewert}@cb.uu.se
[2] Department of Pharmaceutical Biosciences, Division of Toxicology,
Uppsala University, Sweden
{mats.nilsson,bengt.danielsson}@farmbio.uu.se

Abstract. Embryo cultures of rodents is an established technique for monitoring adverse effects of chemicals on embryonic development. The assessment involves determination of the heart rate of the embryo which is usually done visually, a technique which is tedious and error prone. We present a new method for fully automatic heart detection in digital videos of rat embryos. First it detects the heart location, and then it counts the number of heart beats for a predetermined period of time. Using this automated method many more embryos can be evaluated at reasonable cost.

Keywords: Heart Detection, Energy, Embryo, Directional Analysis.

1 Introduction

The technique of whole embryo culture of rodents (where rat and mouse embryos can be cultured outside the uterus) has been used for over four decades to study adverse effects of chemicals and drugs on embryonic development [1]. A major advantage of whole embryo culture is, that development can be directly monitored during organogenesis. For instance, the presence of heart rate and blood circulation, a sign of a healthy embryo, is normally monitored at different times during culture.

Usually this is done manually under a light microscope where the heart rate is determine by simply counting the number of heart beats for a period of time. This can either be done directly by visual inspection, or by filming the embryos and determine the heart rate afterwards from the video recordings by visual inspection. A normal rat heart beats around three times per second. It is quite hard for humans to count heart beats at such high rate, and it even gets worse if one is supposed to count the heart beats for a longer period of time. Instead the heart beats are counted for a period of 15 seconds and then the heart beat rate is calculated from this information.

The manual procedure is tedious and potentially subjective and error prone and there have been some attempts of automating the assessment. In the early

P. Perner and O. Salvetti (Eds.): MDA 2008, LNAI 5108, pp. 27–37, 2008.

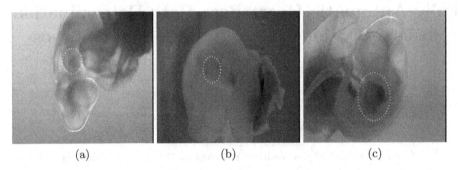

(a)   (b)   (c)

Fig. 1. Examples of some embryo images, where the superimposed white circle represents the heart region. a) An embryo containing lot of variations (topographic structures) in the embryo body, and even the blood color in the heart is hardly visible. b) An embryo with poor contrast and its a situation where the blood color in the heart is not visible. c) An embryo where the blood in the heart is easily visible, but it also contains some dark regions on the left of the heart, which can become good contenders for heart on the basis of blood color.

1970s, a culture system was presented in which embryonic heart rate could be monitored in real-time by the use of a low-intense laser [2]. Although the technique provided the opportunity to monitor the drug effects on the embryonic heart, the authors stated that a complete system represented a considerable investment and that repeated measurements would become expensive.

In 2002 Danielsson et al., presented an automatic method to determine and illustrate heart rate after exposure to drugs with $IKr - blocking$ properties which are known to induce bradycardia, with subsequent arrhythmia, with dose-dependency in cultured rat embryos [3]. Video tapes containing individual embryos are displayed on a monitor and a designed light-sensitive probe is hand-held on the pulsating embryonic heart. The resulting analogue signal is A/D converted and displayed as a trace showing the difference in light intensity when the embryonic heart is filled and emptied of blood. This trace can then be used for determining heart rate [3], [4].

Although the methods mentioned above estimate the heart frequency to a satisfactory level, they require rather complicated equipment. In this paper we present a new method which only requires a standard digital video camera attached to a microscope and a normal PC for the analysis.

For validation of our method, videos were stepped through frame by frame and the exact heart rate was determined visually. This very tedious procedure was used in order to obtain a *gold standard* or ground truth for comparison with the other methods. Determining the heart rate through image analysis of the whole image field turns out not to work because of too many disturbing structures and motions. Some of the problems are easily visible from Fig. 1. We thus need to automatically find the heart location so that the analysis can be focused there. The automated detection of heart rate will involve four steps: 1. Finding the embryo border 2. Finding the heart within the embryo. 3. Tracking the blood flow back and forth in the beating heart so that a time signal expressing the heart beat is obtained. 4. Analyzing the time signal to determine the heart rate and its variations.

2   Finding the Embryo Border

The most prominent feature in the image sequences that can be used to detect the heart location is the fact that the blood as well as the heart structures is seen pulsating at the heart rate. Direct application of motion detection filters, however, fails because the embryo as a whole also moves significantly, both with the heart beats and also for other reasons. We thus need to separate the motion of the embryo as a whole from the motion of the heart. The heart boundary is rather diffuse and has low contrast, while the embryo boundary is sharper and has a much stronger contrast. Due to the smooth nature of the heart boundary, the spatial frequencies representing the heart boundary will thus be lower than those of the embryo boundary. The stronger contrast for the embryo boundary also makes it easier to detect its gradient directions. This means that if we employ some kind of directional analysis tool, it will separate the embryo boundary from the rest of the embryo.

Directional analysis can be either local or global. Sobel, Prewitt, Laplacian etc., [5] are normally used for local directional analysis. All these filters provide good edges but are limited to local orientation due to their limited size. If the edge width is changing, you need to apply varying size filter to detect the edges. Gabor filters [6] also provide directional analysis for particular frequency and orientation, but is local in nature. Decimation-Free Directional Filter Banks (DDFB), provides global directional analysis. They split the spectrum of an image into wedge shaped passbands [7], and in the spatial domain these passbands correspond to information in a particular direction, irrespective of whether it has a low or high frequency content. So, global directional analysis is not sensitive to local orientation, and hence less sensitive to noise.

Visually it seems as if a threshold would separate the embryo from the background but severe shading causes this to fail as illustrated in Fig. 2c where we have applied thresholding followed by simple boundary detection on Fig. 2a. Local filters will not work well either, for instance applying a Laplacian gives the result shown in Fig. 4b. Under the very same circumstances, DDFB has extracted the boundary very accurately as shown in Fig. 2d, mainly due to its less sensitive nature to noise and local orientation. Here we have used DDFB which partitions the full spectrum of an image into eight equal partitions [7]. Fig. 3a, and b shows two outputs of DDFB when applied to Fig. 2a.

The computational complexity would be really high if we apply DDFB on each image in the video sequence. Instead of applying DDFB to the whole video sequence, we have applied it on the first three frames. But obviously the motion of the whole embryo boundary cannot be captured with only the first three frames of the sequence. The motion of the embryo is random but it is not translating in any direction. This random motion can be masked by energy calculation. Here

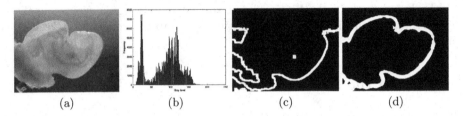

(a)   (b)   (c)   (d)

Fig. 2. Example of boundary detection. a) Image having varying illumination and contrast. b) Histogram of (a). c) *Otsu's* thresholding of (a). Here the boundary is dilated for better visualization. d) Boundary detection by applying DDFB on (a). Once again the boundary is dilated for better visualization.

energy is calculated in the same manner as described in [7], but we have used it in a new way. As a background, we will give a short description of how energy is used in DDFB for image enhancement.

The outputs of DDFB are normally named *directional images*. Energy is calculated for each of these *directional images* which result in corresponding *energy images*. To get a value for any position (a, b) in the *enhanced image*, the energy values at position (a, b) in all *energy images* are compared. Then for the *energy image* containing the highest energy value, the value from its corresponding *directional image* is picked, and placed at position (a, b) in the *enhanced image*. This process will find any edges irrespective of their sharpness in the image, and try to make them part of enhanced image. So, whenever DDFB is used for image enhancement, energy is used only for comparison purpose, but its value is not part of the enhanced image. But here we are interested in only the embryo boundary, and we know that it has got good contrast.

In the *energy images*, the region containing the boundary of the embryo will be represented by higher values, while the rest of the *energy image* should have a darker look due to weak edges. Fig. 3c, and d shows two *energy images* resulting from Fig. 3a, and b respectively. It is evident that the higher energy values correspond to edges in their respective *directional images*. So, a simple segmentation on these *energy images* will provide us with the embryo boundary along with some extra isolated structures. Fig. 3e, and f shows two segmented images resulting from Fig. 3c, and d respectively. A simple *logical or* operation between these *segmented energy images* will provide us with the whole embryo boundary. Here we have to select the largest connected component in the resultant image to remove the small isolated structures, which are normally produced due to noise, and parts of the embryo which are almost as sharp as the embryo boundary.

This new way of segmentation can also be used to differentiate between strong edges and relatively weak edges. This can be achieved by just masking the strong edges found in the original image, and then repeating the same process as mentioned earlier for strong edges, to find the rest of the edges. It is computationally more efficient then the previous algorithm where segmentation is done

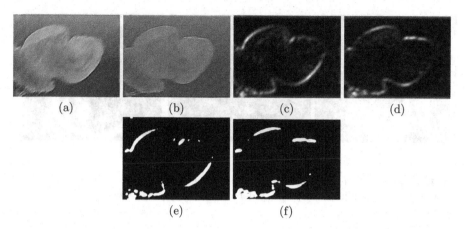

Fig. 3. Outputs of DDFB when applied to Fig. 2a. Here a) and b) shows two out of eight *directional images* when DDFB is applied to Fig. 2a. c) *Energy image* of (a). Higher values in *energy image* represents the edges. d) *Energy image* of (b). e) *Otsu's* thresholding of (a). f) *Otsu's* thresholding of (b).

after image enhancement. To compute *energy images*, we have used the formula shown in equation (2):

$$f_s(x, y, n) = f(x, y, n) - \frac{1}{m^2} \sum_{c=\frac{-m+1}{2}}^{\frac{m-1}{2}} \sum_{d=\frac{-m+1}{2}}^{\frac{m-1}{2}} f(x - c, y - d, n), \qquad (1)$$

$$f_e(x, y, n) = \sum_{c=\frac{-m+1}{2}}^{\frac{m-1}{2}} \sum_{d=\frac{-m+1}{2}}^{\frac{m-1}{2}} | f_s(x - c, y - d, n) | . \qquad (2)$$

Here, $f(x,y,n)$ represents a time frame sequence. x, and y represent the spatial horizontal and vertical dimensions respectively. n represents the frame number. In equation (2), $f_e(x, y, n)$ represents the *energy image*. m, the block size, is the most critical parameter. It determines how large motions of the embryo wall that can be handled. We are assuming that the embryo boundary is moving randomly, but the overall embryo is not translating in any direction. Thus, using $m = 30$ is a good choice. To completely detect the boundary, we pass each of the first three time frames through DDFB, which will result in eight directional images. On each directional image, equation (1), and (2), is applied which results in eight *energy images* for each frame. On each of these *energy images* automated thresholding based on *Otsu's* method is applied, resulting in a total of 24 binary images. All these images are logically *or* to form a *template image*. Then only the biggest connected component in *template image* is preserved. This *template* is used to mask out the embryo boundary for the subsequent analysis steps. We represent the boundary masked time frame sequence by $f_m(x, y, n)$, where x, y, and n represent the same dimensions as earlier.

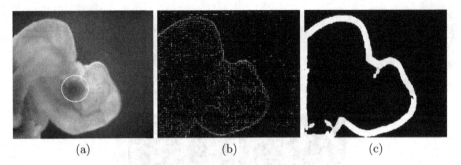

(a)   (b)   (c)

Fig. 4. Thresholding results. a) Ist frame of a time frame sequence. b) Laplacian of (a) followed by Thresholding. c) Boundary detection by applying DDFB on (a).

3   Finding the Heart

Once the movement of the boundary is masked out, the remaining dominant movement would be due to the heart motion, but there are still sampling and noise problems which prevents the simple solution of taking the standard deviation across the time frame sequence to find the heart location. Instead we have used an energy based algorithm for this purpose.

Lets think of the boundary masked time frame sequence as a *volume image*, with the two spatial dimensions(x, y) and one time dimension (n). Now, we slice this *volume image* into two dimensional images where either one of the spatial dimension and the time dimension will serve as the two dimensions of the *sliced image*. Mathematically it can be represented as,

$$g_x(y, n) = f_m(x, :, :), \qquad (3)$$

where $g_x(y, n)$, represents the *sliced image*, along row number x, and ":" represents the set of all possible values for the dimensions y, and n. To avoid the effect of non-uniform illumination, we have used a butterworth filter as discussed in [7] on each *sliced image* $g_x(y, n)$.

Equation (3) will be repeated for each value of x, which will result in as many images as the number of rows in the *volume image*. The next step is to identify those rows where the heart is located. To accomplish this, we take the Fourier Transform of the *sliced images*. Mathematically it can be represented by,

$$G_x(u, v) = \sum_{n=0}^{c-1} \sum_{y=0}^{b-1} f_m(x, y, n) e^{-j2\pi(\frac{uy}{b} + \frac{vn}{c})}, \qquad (4)$$

where b denotes the total number of columns, and c denotes the total number of frames, and u, v can take values $0, 1, 2, ..., b-1$, and $0, 1, 2, ..., c-1$ respectively.

Fig. 5a) shows a *sliced image* when we are out of the heart region. Fig. 5b) shows a *sliced image*, when we are in the heart region. One can easily see the periodic pattern in Fig. 5b) which represents the motion of the heart. This

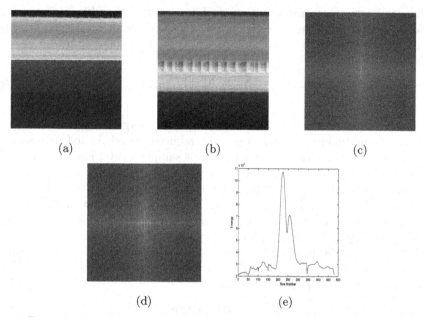

(a)   (b)   (c)

(d)   (e)

Fig. 5. Example of some image slices and their Fourier Transform, of the image shown in Fig. 2a. a) Slice of the time frame sequence when we are out of the heart region. Clearly, there is no sort of periodic pattern resembling the heart motion in the image. b) Slice of the time frame sequence when we are in heart region. This periodic pattern is produced due to the heart beat. c) Magnitude Fourier Transform of the Image shown in (a). As there was no sort of periodic pattern in (a), the transformed image has almost no activity. d) Magnitude Fourier Transform of the image shown in (b). All the periodic activity in (b) is represented in its Fourier Transform by the pattern which appears along the horizontal axis of the magnitude transform. e) The plot shows the energy of the Fourier Transform $R(x)$.

has motivated us to model the motion of a heart as a *sinusoid*. Since we have knowledge about how fast a heart can beat, we can use this information in the development of our method. To use this information we are assuming that the heart will not beat more then 6 times per second. Currently, the sequence is recorded at 25 frames per second. Which means that the highest frequency representable is 25/2 (Shanon Sampling Theorem), which would appear on π, and $-\pi$ in the Fourier Transform of $G_x(u,v)$ on u axis, but we cannot say anything about v axis. So, our heart beat frequency will always reside between $-\pi/2$ to $\pi/2$. For this reason we have computed the energy on u axis of $G_x(u,v)$ between this range using all possible values of v. We do not need to include the rest of the frequency range as this is not representing the heart motion. As we have already filtered these *sliced images* with a butterworth highpass filter, which means that the DC component and certain low frequencies around DC component of the Fourier Transform will not contribute to the energy calculation.

Now for each row number x, we have a Fourier Transform $G_x(u,v)$. Fig. 5(c) shows the magnitude Fourier Transform of image shown in Fig. 5(a). Here the magnitude Fourier Transform contain much less activity than Fig. 5(d) which is a Fourier Transform of the *sliced image* shown in Fig. 5(b). The higher activity in Fig. 5(d) represents the heart motion. This process is followed by,

$$R(x) = \sum_{u=(c/4)+1}^{(c*3)/4} \sum_{v=0}^{b-1} | G_x(u, v) |, \qquad (5)$$

where $R(x)$ represents a one dimensional vector of energies. One can easily see the outer summation range in equation (5) is adjusted to only include frequencies between $-\pi/2$ to $\pi/2$ according to the current sampling rate. Fig. 5e shows the Fourier energy plot. Here the higher energy values represent the rows where the heart is located. The repetition of the same procedure from equation (3) to equation (5), just replacing x with the y variable will result in $R(y)$ which finds the columns containing the heart. This process is followed by equation (6), which is basically an image f_{img}, formed by using both the energy vectors. The application of *Otsu'a* thresholding method on f_{img} results in a binarized image, where the non-zero area in f_{img} *thresholded image* represents the heart location in each frame of the time frame sequence.

$$f_{img} = R(x)' * R(y), \qquad (6)$$

where, $R(x)$, $R(y)$ are energy row vectors as discussed earlier, and $'$ represents the transpose operation. We will use the indices of non-zero values of f_{img} *thresholded image* to crop the heart from each frame of the time frame sequence, and named it the *heart image*. Fig. 6a shows the heart location outlined over the image shown in Fig. 2a.

4   Creating a One Dimensional Signal Representing the Heart Beat

After finding the heart position in the time frame sequence, the next step is to extract a signal that represents the heart beat. We have used 2-D cross-correlation between a template and the *heart image* sequence to generate this signal.

The template should represent either of the extremes states in the heart cycle, either end-systole or end-diastole. End-systole represents that the heart is in contracted position, driving blood out of the heart chambers, and end-diastole is its opposite. One can think of one heart beat time as the time it takes between two consecutive either end-systole or end-diastole. To find such a template we create a similarity matrix from the first Z frames of the video sequence, representing video from the first 2 seconds. The dimension of the matrix is thus $Z \times Z$. Each entry of this matrix can be calculated as,

$$S_{ij} = A(H_i, H_j), \qquad (7)$$

where S_{ij} represents the entry at row i, and column j. H_i, H_j represent the *heart image* from frame i, and j respectively. Here A represents the cross correlation which can be represented by the following formula.

$$A(p,q) = \frac{\sum_{x=0}^{a-1}\sum_{y=0}^{b-1}(p(x,y)-\bar{p})(q(x,y)-\bar{q})}{\sqrt{\sum_{x=0}^{a-1}\sum_{y=0}^{b-1}(p(x,y)-\bar{p})^2 \sum_{x=0}^{a-1}\sum_{y=0}^{b-1}(q(x,y)-\bar{q})^2}} \tag{8}$$

here a, b represent the total number of rows and columns of image p and q. $\bar{p}$, $\bar{q}$ represent the mean of the image p and q.

The minimum value in this matrix represents the worst cross-correlation, which means that the heart is in opposite *state* in the two images whose cross-correlation is measured. The location of the minimum value in the similarity matrix will provide us with the frame number for the corresponding images. We choose one of the frames corresponding to the minimum correlation entry as our *heart template image*, H_T.

Now we are in a position to find the one-dimensional function (which evaluates with time) representing the heart motion. It is created simply by taking the cross-correlation between each heart image and the *heart template image*. Mathematically it can be represented as,

$$H_{graph}(n) = A(H_T, H_n), \tag{9}$$

where H_n represents the heart image, and n represents the frame number. Fig. 6b shows a graph showing the heart beat motion as a function of time(frame number). It clearly shows a periodic pattern which represents the heart beat, and is *sinusoidal* in nature, which justifies our statement in the last section.

5   Determining the Heart Frequency

From the signal representing the heart beat, we simply need to find all significant local maxima or all significant local minima in H_{graph}. The total number of local minima or maxima will represent the total number of heart beats for a selected time period typically 15 seconds. We can also analyze the variation in this signal to see major deviations in the heart rhythm indicating serious arrhythmias.

6   Experimental Results

We have tested our method on a set of 30 cases. All these videos are taken from different days during the culturing process to include the shape variations of embryos, embryo boundary sharpness variations, and also the change in recording environment. The time frame sequences are stored in AVI format with 480x640 pixels per frame. With 25 frames per second, a 25 second long time frame sequence will occupy around 183 Mega-Bytes when stored as unsigned 8-bit integer. Using Linux based $MATLAB$(7.40.287) with $Intel(R)$ $Xeon(TM)$ 3.6GHz processor, It takes 120 seconds to complete the analysis, out of which half of the time is taken for loading the video into the RAM.

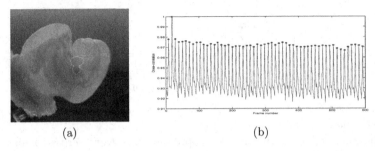

(a)   (b)

Fig. 6. Heart location and plot of heart movement. a) The heart location in Fig. 2a found from 'f_{img} *thresholded image*' is outlined. b) The plot show the heart movement as a function of frame number.

Our method was able to detect the embryo boundary, heart location and determine the heart beat automatically in all 30 cases. To verify the accuracy we have compared our results with the results from stepping through the image sequences frame by frame and manually determining the number of beats, thus creating a *gold standard*. We have also compared our results with those obtained with the previously used manual method where the heart beat is counted in real time. Keeping up with one to two beats per second is difficult as seen from the results in Table. 1. The *2nd* row shows the manually counted results, while the *3rd* row shows the result by applying our method and the *4th* row shows the *gold standard*. From the table, it is clear that our results are within ±1 of the results shown in row 4, whereas the variations in manually counted results are slightly higher.

Table 1. The table shows the result of 20 time frame sequences calculated over the time of 15 Seconds. Here M, P, S, stands for Manual, Proposed, and Slow motion respectively.

	V1	V2	V3	V4	V5	V6	V7	V8	V9	V10	V11	V12	V13	V14	V15	V16	V17	V18	V19	V20
M	42	45	45	43	42	41	36	33	35	40	45	38	36	39	35	41	39	43	43	42
P	41	45	47	46	41	42	35	33	36	39	46	40	36	41	35	43	43	45	47	42
S	40	44	46	47	41	42	35	33	36	39	46	40	36	42	34	43	43	46	48	42

7   Conclusion and Future Work

We have presented a fully automated approach to detect and count the heart beat in rat embryos. We have used the DDFB in a new way, so that strong edges can be segmented directly by skipping the enhancement stage. The method may seem rather complex but our attempts to use simpler approaches have resulted in significantly less robust solutions, mainly due to large variations in shape of embryo during the culturing process, and digitization noise. There is also movement from the liquid culture medium, which is produced due to the movement of the heart. The method will fail whenever there is a translation in embryo. However, this can be solved by updating the *heart template image* by using equation (3), whenever there is drop in cross-correlation value, but it would be computationally much expensive.

In our present method the heart rate is robustly determined and significant deviations in the inter beat intervals can be detected and reported as an indication of arrhythmia. The detection of arrhythmia is, however, a very important aspect of this kind of analysis and we are currently exploring more sophisticated analysis of the heart beat signal to detect unusual patterns that can be indicative of arrhythmia.

Acknowledgements

The authors would like to thank Raili Engdal for the technical support. The authors also acknowledge Swedish Medical Product Agency and CIIT Islamabad, for the financial support.

References

1. Flick, B., Klug, S.: Whole Embryo Culture: An Important Tool in Developmental Toxicology Today. Current Pharmaceutical Design 12, 1467–1488 (2006)
2. Robkin, M.A., Shepard, T.H., Tanimura, T.: A New in Vitro Culture Technique for Rat Embryos. Teratology 5, 367–376 (1972)
3. Sköld, A.C., Danielsson, C., Linder, B., Danielsson, B.R.: Teratogenicity of the IKr-Blocker Cisapride: Relation to Embryonic Cardiac Arrhythmia. Reproductive Toxicology 16, 333–342 (2002)
4. Danielsson, C., Azarbayjani, F., Sköld, A.C., Sjögren, K., Danielsson, B.R.: Polytherapy with hERG-blocking Antiepileptic Drugs: Increased Risk for Embryonic Cardiac Arrhythmia and Teratogenicity. Birth Defects Research Part A: Clinical and Molecular Teratology 79, 595–603 (2007)
5. Umbaugh, S.E.: Computer Vision and Image Processing: A Practical Approach Using CVIP Tool. Prentice Hall International, New Jersey (1998)
6. Dunn, D., Higgins, W.E.: Optimal Gabor Filters for Texture Segmentation. IEEE Transactions on Image Processing 4, 947–964 (1995)
7. Khan, M.A.U., Khan, M.K., Khan, M.A.: Coronary Angiogram Image Enhancement using Decimation-free Directional Filter Banks. In: IEEE International Conference on Acoustics, Speech, and Signal Processing, pp. 441–444. IEEE Press, Montreal (2004)

Biomedical Signal and Image Processing for Decision Support in Heart Failure

Franco Chiarugi[1], Sara Colantonio[2], Dimitra Emmanouilidou[1],
Davide Moroni[2], and Ovidio Salvetti[2]

[1] Institute of Computer Science (ICS)
Foundation for Research and Technology Hellas (FORTH)
Heraklion, Crete, Greece
{chiarugi,dimeman}@ics.forth.gr
[2] Institute of Information Science and Technologies (ISTI)
Italian National Research Council (CNR), Pisa, Italy
{sara.colantonio,davide.moroni,ovidio.salvetti}@isti.cnr.it

Abstract. Signal and imaging investigations are currently a basic step of the diagnostic, prognostic and follow-up processes of heart diseases. Besides, the need of a more efficient, cost-effective and personalized care has lead nowadays to a renaissance of clinical decision support systems (CDSS).

The purpose of this paper is to present an effective way to achieve a high-level integration of signal and image processing methods in the general process of care, by means of a clinical decision support system, and to discuss the advantages of such an approach.

Among several heart diseases, we treat heart failure, that for its complexity highlights best the benefits of this integration.

Architectural details of the related components of the CDSS are provided with special attention to their seamless integration in the general IT infrastructure. In particular, significant and suitably designed image and signal processing algorithms are introduced to objectively and reliably evaluate important features that, in collaboration with the CDSS, can facilitate decisional problems in the heart failure domain. Furthermore, additional signal and image processing tools enrich the *model base* of the CDSS.

1 Introduction

Signal and imaging investigations are currently a basic step of the diagnostic, prognostic and follow-up processes of heart diseases. Not by chance, in the last decades, the development of Computer-Aided Diagnosis (CAD) schemes has attracted a lot of interest and effort within medical imaging and diagnostic radiology, becoming in some cases a practical clinical approach. The basic concept of CAD is to provide a *second opinion* or a *second reader* that can assist clinicians by improving the accuracy and consistency of image based diagnoses [1]. Actually, the clinical interpretation of diagnostic data and their findings largely depends on the reader's subjective point of view, knowledge and experience.

P. Perner and O. Salvetti (Eds.): MDA 2008, LNAI 5108, pp. 38–51, 2008.

Hence, computer-aided methods, able to make this interpretation reproducible and consistent, are fundamental for reducing subjectivity while increasing the accuracy in diagnosis. As such, they are likely to become an essential component of applications designed to support physicians' decision making in their clinical routine workflow. Other important motivations rely on the limits to reader's ability of data interpretation caused by either the presence of structure noise or the vast amount of data, generated by some devices, which can make the detection of potential diseases a burdensome task and may cause oversight errors.

Besides, the development of computerized applications for supporting health care givers (an old but still alive quest, started more than 45 years ago in the early 1960s) is experiencing a period of rapid expansion in knowledge, motivated by a renewed interest [2]. The need of a more efficient, cost-effective and personalized care and of a more rational deployment of diagnostic resources is one of the reasons behind this renaissance. Actually, the development and increasing use of hospital or, even, cross-enterprise regional health information systems make possible the design of ambitious integrated platforms of services in order to guarantee the continuity of care across the various stakeholders. Clinical Decision Support Systems (CDSSs) are a natural and key ingredient of such integrated platforms, since they may compete with the increasing bulk of clinical data by providing an integrated approach to their analysis. In addition, CDSSs may foster adherence to guidelines, prevent omissions and mistakes, spread up-to-date specialistic knowledge to general practitioners and so on.

This being the general setting, the purpose of this paper is to address the integration of signal and imaging investigations with the wide-ranging services provided by CDSSs. Actually, signal and image processing methods may be understood and embedded as a part of the *model base* of the CDSS. In such a way an effective high-level integration of signal and image processing methods in the general process of care is achieved.

With the aim of avoiding unnecessary generality, the paper addresses the specific yet complex and paradigmatic example of image and signal processing for decision support in *heart failure*. Indeed, heart failure is a clinical syndrome, whose management requires –from the basic diagnostic workup– the intervention of several stakeholders and the exploitation of various imaging and non-imaging diagnostic resources.

The paper is organized as follows. First, heart failure management is briefly described in Section 2.1, including a description of its diagnostic workup which is enlightening to understand the complexity of this syndrome. In Section 2.2 the quest for a decision support system is motivated, describing relevant decisional problems. In Section 3, signal and imaging investigations are justified, highlighting the value added to the CDSS, while suitably designed algorithms for image and signal processing are introduced in Sections 3.2 and 3.3 respectively. In Section 4, the results of architectural design for integration are described both at the IT infrastructure level (Section 4.1) and at the higher level represented by the general CDSS (Section 4.2). Finally, Section 5 ends the paper with some remarks and directions for future work.

2   Background

2.1   Heart Failure

Heart Failure (HF) is a complex clinical syndrome resulting from any structural or functional cardiac disorder which impairs the ability of the ventricle to fill with or eject blood. In its chronic form, HF is one of the most remarkable health problems for prevalence and morbidity, especially in the developed western countries, with a strong impact in terms of social and economic effects. All these aspects are typically emphasized within the elderly population, with very frequent hospital admissions and a significant increase of medical costs.

The first, immediate and enlightening proof of HF complexity is represented by its diagnostic workup, which we briefly describe next. Indeed, it can be considered as the first stage of HF patients' management which necessarily requires the acquisition and analysis of signal and imaging data.

Heart Failure Diagnostic Workup. Figure 1 shows the sequence of steps that compose the HF diagnostic workflow [3]: after having assessed the presence of main signs and symptoms, physicians usually require diagnostic examinations such as ECG, chest X-ray and neuroendocrine evaluations (i.e. Brain Natriuretic Peptides - BNP) in order to check out the diagnosis, confirmed eventually by an echocardiography investigation. Supporting such a decision problem requires to encode the workflow into an opportune knowledge base which formalizes, for each step, the set of conditions evaluated by physicians. The first step regards the presence and severity of signs and symptoms such as breathlessness, swelling, fatigue, hepatomegaly, elevated jugular venous pressure, tachycardia, third heart sound and pulmonary crepitations. Then, ECG signals are acquired for investigating the presence of anterior Q waves and left bundle branch block, signs of left atrial overload or left ventricular hypertrophy, atrial fibrillation or flutter and ventricular arrhythmia. If ECG abnormalities are present, HF diagnosis is considered carefully possible and further checked out by analyzing chest X-ray. Such an examination is useful for detecting the presence of cardiac enlargement (cardio-thoracic ratio > 0.50) and pulmonary congestion. In parallel, neuroendocrine analysis are performed to test out high levels of natriuretic peptides which suggest the presence of a cardiac disease. Whether all these examinations certify the presence of abnormalities, an echocardiographic investigation is performed for documenting a cardiac dysfunction. The most important parameter to be evaluated from such a diagnostic modality is the Left Ventricle Ejection Fraction (LVEF); other relevant data are the fractional shortening, the sphericity index, the atrioventricular plane displacement, the myocardial performance index, the left ventricular wall motion index, the isovolumic relaxation time, the early to atrial left ventricular filling ratio, the early left ventricular filling deceleration time, the pulmonary venous atrial flow velocity duration, the ratio of pulmonary vein systolic and diastolic flow velocities, and the pulmonary artery pressures. HF diagnosis is finally concluded if symptoms and signs and ECG / X-ray / BNP level / Echocardiographic abnormalities are all present.

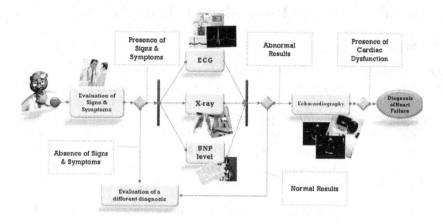

Fig. 1. HF Diagnostic Workflow

2.2   Decision Support in Heart Failure

Recent studies and experiences have demonstrated that accurate heart failure management programs, based on a suitable integration of inpatient and outpatient clinical procedures, might prevent and reduce hospital admissions, improving clinical status and reducing costs. Actually, HF routine practice presents several aspects in which an automatic, computer-based support could have a favourable impact. A careful investigation about the needs of HF practitioners and the effective benefits assured by decision support was performed: four problems have been identified as highly beneficial of CDSS point-of-care intervention [4]. They can be referred as macro domain problems and listed up as: (i) HF diagnosis, (ii) prognosis, (iii) therapy planning, and (iv) follow-up. Further detailed decision problems were identified for specifying these macro domains, focusing as much as possible on the medical users' needs; explicative examples are:

- severity evaluation of heart failure
- identification of suitable pathways
- planning of adequate, patient's specific therapy
- analysis of diagnostic examinations
- early detection of patient's decompensation

An accurate analysis highlighted that the needed corpus of knowledge mainly consisted of domain know-how. Nevertheless, the solution of some of these problems seemed still debated in the medical community, due to the lack of validated and assessed evidences. In such cases, computational models appeared the best solution for modelling the decision making, extracting knowledge directly from available data.

In this perspective, a CDSS for the management of heart failure, which combines several models of reasoning, has been suitably designed. Having the overall organization of the CDSS being reported in [5], the focus in the sections below is on the analysis of diagnostic examinations and on their integration into the CDSS.

3   Signal and Image Processing in Heart Failure

3.1   Significance

During the formalization of the main decisional problems that require the CDSS intervention and, hence, listing up all the pieces of knowledge, data and information relevant for decision making, the importance of considering and interpreting ECG signals and echocardiography images had come forth. Indeed HF diagnostic workup was a straightforward example of the importance of computer-aided data processing in HF decision making, but other significant contributions can be envisaged. Overall, among all the profitable applications into decision support workflows, the following can be listed up:

- automatic or semi-automatic computation of parameters relevant in the decisional problems;
- support of physicians' case-based reasoning processes;
- discovery of novel pertinent knowledge.

While the first is typical of routine workflows in relatively simple situations, as described in the diagnostic workup example, the other two can be considered advanced applications that may aid physicians in facing critical cases or critical problems.

Actually, not only the parameters extracted from signals and images examinations are significant to physicians for formulating a response but also the data themselves can be useful for having a general overlook of a patient's situation. This means that allowing clinicians to explore data can assure the availability of a lot of other pieces of information hidden in the same data. Moreover, when dealing with a difficult case, comparing the one at hand with assessed responses for other patients' situations can be really helpful [6]. This entails maintaining and making available a database of cases with annotated images and signals which can be retrieved by similarity on a set of computed features (see Section 4.1). Difficult diagnoses and, most of all, prognosis assessment are examples of these situations. For such critical problems, data processing facilities can have further relevance for the discovery of novel knowledge by granting the computation of a wide range of parameters which can be explored and correlated in order to find out new relevant patterns [7].

Finally, from the opposite side, opportune knowledge formalization may represent advantages in personalization of diagnostic imaging and non-imaging investigations. This means that adequate conditions could be encoded within the CDSS in order to suggest which kind of parameters could be more usefully evaluated for a given patient during, for instance, an echocardiography or an ECG session.

3.2   Image Processing Methods

Imaging techniques offer invaluable aid in the objective documentation of cardiac function, allowing for the computation of parameters relative to chamber dimensions, wall thickness, systolic and diastolic function, regurgitations and pulmonary blood pressure. As previously mentioned, chest X-ray and echocardiography should be included in the HF initial diagnostic workup. Further, echocardiography will be regularly repeated to monitor in an objective way the changes in the clinical course of a HF patient. Additional techniques, like nuclear imaging and cardiac magnetic resonance, may be also considered for particular patients, since they have not been shown to be superior to echocardiography in the management of most HF population. Thus, echocardiography and in particular 2-D TransThoracic Echocardiography (TTE) for its portability and versatility is the key imaging technique for the practical management of HF. The most important measurement performed by TTE is LVEF, which permits to distinguish patients with cardiac systolic dysfunction from patients with preserved systolic function. LVEF is given by the normalized (non-dimensional) difference between left ventricle End-Diastolic Volume (EDV) and the End-Systolic volume (ESV). Among different models for the computation of such volumes, the American Society of Echocardiography [8] suggests the use of the so-called Simpson's rule, by which the left ventricle is approximated by a stack of circular (or elliptical) disks whose centers lie in the major axis. Simpson's method, therefore, relies on left ventricle border tracing. It is well-known that manual border tracing, besides being time-consuming, is prone to inter- and intra- observer variability, and thus is unable to provide a satisfactory and reproducible measurement of LVEF. Image processing techniques may reduce variability of human interventions in border tracing, by providing automated or, at least, semiautomated methods for tracing contours of relevant structures found in an image. However, the segmentation problem for ultrasound images is by no means trivial, due mainly to low signal to noise ratio, low contrast, image anisotropy and speckle noise [9]. Nevertheless, some acquisition devices already offer the possibility of automatically computing a set of relevant parameters but are still really expensive and this is the reason why older devices are still very common.

From these considerations, it was early realized that the development of assisted segmentation methods, able to deal with echocardiographic image sequences, could represent a valid support to the physicians in the process of image report formation.

Thus a prototypical toolkit [10] –composed of three main modules– for the analysis of apical-view sequences of the heart has been developed. Two typical frames of such sequences are shown in Figure 2.

The first module (*Region Identification*), which takes in input an apical sequence of the heart, is able to identify the left ventricle cavity in every frame of the sequence by means of *mimetic criteria*, providing a rough segmentation.

The second module (*Segmentation Refinement*), which takes in input an image and a rough segmentation of it, is able to refine the segmentation exploiting a variational formulation of level set methods, which achieves regularization of the boundary of the left ventricle as well as better adherence to image edges [11].

The third module (*Feature Extraction*) is able to extract significant features from a set of segmented left ventricles, the most important being EDV and ESV (both computed according to Simpson's rule) and, in turn, LVEF.

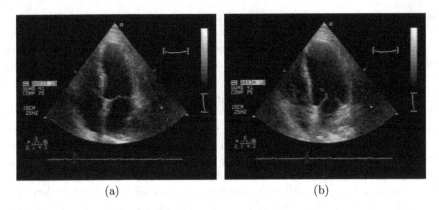

(a)   (b)

Fig. 2. Typical frames of an image sequence taken from the apical view

After the integration in a suitable graphical user interface, three possible ways may be foreseen to employ the toolkit. These ways are described below according to the automatism level, starting from the less automatic one.

Case A) Manual Selection of the End-Diastolic and End-Systolic Frames and Rough Manual Contour Tracing. In this case, the toolkit provides a refinement of the manually traced left ventricle contour in the manually selected frames. Instead of using the common free hand selection, the user may just quickly select a polygonal region approximating the left ventricle cavity. The *Segmentation Refinement* module is then triggered. In particular, the manually drawn contour is used for the initialization of the level set method. Finally, the third module is used for feature extraction.

Case B) Manual Selection of the End-Diastolic and End-Systolic Frames and Automatic Contour Tracing. In this case, the toolkit traces automatically the contour of the left ventricle in the manually selected frames. The *Region Identification* module is used to find an approximate left ventricle contour. Then the contour is refined by the level set segmentation step as in Case A).

Case C) Automatic Selection of the End-Diastolic and End-Systolic Frames and Automatic Contour Tracing. In this case the toolkit takes in input the whole image sequence and applies the *Region Identification* module to every frame in order to obtain a rough segmentation of the left ventricle. Then the volume of the cavity is computed on this rough segmentation by using the *Feature Extraction*. The indices of the frames corresponding to the extremal values (i.e. maximum and minimum) of the volume are found and stored. Then, the *Segmentation Refinement* is applied to the contours in the frames which are near to those of extremal values. Computing again volumes on the basis of the refined contours by the *Feature Extraction* module leads to the identification of the end-systole and end-diastole frames and to the computation of related clinical parameters. The final result of segmentation in the automatically identified end-systole and end-diastole frames is shown in Figure 3.

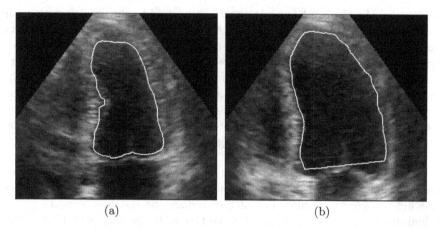

(a)   (b)

Fig. 3. Final result of segmentation in an end-systole (a) and in an end-diastole (b) frame

The proposed image processing toolkit could be easily extended in several ways. Besides integrating standard tools for performing graphically image measurements (such as linear measurements) and producing IHE-compliant Simple Image and Numeric Reports, the core segmentation modules may be adapted to deal with other echocardiographic views, so as to perform a complete quantification of heart chambers.

3.3   Signal Processing Methods

ECG is one of the very basic examinations performed in the evaluation and assessment of HF. According to [3], the negative predictive value of normal ECG to exclude left ventricular systolic dysfunction exceeds 90%. The most common ECG examinations are the *Resting ECG* and the *Holter ECG*. While the latter is more commonly used for the discovery of rhythm abnormalities and the computation of the Heart Rate Variability (HRV), the former is more commonly used for the evaluation of morphological abnormalities in the PQRST shape.

Considering the crucial role of ECG signals and the various related examinations, it has been immediately judged important to design and implement some *basic*, *robust* and *scalable* algorithms for ECG processing that could be immediately applied to the raw data acquired by ECG devices with different lead numbers and different acquisition periods. After some interviews with the clinicians, it has been identified a significant operative scenario, where the ECG acquired with a non-interpretive electrocardiograph is transferred to the hospital gateway and from there processed in order to:

1. Detect the QRS complexes
2. Identify the dominant beats
3. Evaluate the averaged dominant beat (for all the leads)

In particular, the averaged dominant beat can be used by the cardiologists (with the help of a graphical ECG viewer), for the evaluation of all the measurements of interest for the diagnosis or the follow-up of heart failure patients, like ST depression, QRS and QT durations, Sokolow-Lyon index for left ventricular hypertrophy, presence of left or right branch bundle block and presence of pathological Q waves. Notice that, since the average dominant beat is cleaner from the noise than the original signal, performing measurements on this average beat leads to a more accurate results, thus reducing inter- and intra- observer variability. The algorithms developed for ECG processing are briefly described below.

QRS Detection. The selected approach for QRS detection belongs to the time-domain techniques [12]. The first step consists in a signal pre-filtering using a moving-average linear filter in order to reduce the baseline wandering and the high-frequency noise, and to select the typical frequencies contained in the QRS complexes. Then a QRS enhanced Signal (QeS) is built as the sum of the absolute derivatives of each pre-filtered channel. The filter for the generation of the derivatives has been chosen trying to reduce the effect of the high frequency residual noise. In practice a pass-band filter is used with a derivative behavior in the band of interest. Then, the beginning of a QRS is detected when the QeS overcomes a suitably defined adaptive threshold. Using only the above algorithm the QRS detection results are good enough, especially in recordings with low or medium content of noise. However, when the noise in one or both leads is high, the performances of the detector are significantly reduced. Therefore, a technique for the improvement of the detection performance when the noise is present only in one channel has been introduced. In particular a Noise Index (NI) is associated with every detected QRS on the basis of the T-P interval average power divided by the QRS average power [13].

Since the NI can be used as an indicator of the noise in the two different channels and of good QRS detection, the appearance of a number of consecutive noisy QRSs determines the beginning of a noisy interval, which ends once a few consecutive non-noisy QRSs appear. In this way, a procedure for best channel selection can be obtained with significant improvement of the overall QRS detection performance. The results have been evaluated on the 48 records of the MIT-BIH Arrhythmia Database where each ECG record is composed by 2 leads sampled at 360 Hz for a total duration of about 30 minutes. The annotated QRSs are 109494 in total. The results have been very satisfying on all the annotated QRSs and, with the inclusion of an automatic criterion for ventricular flutter detection, a sensitivity=99.76% and a positive predictive value=99.81% have been obtained.

Construction of the Average Dominant Heart Beat. A prerequisite for the construction of the average dominant beat is the morphological classification of each detected QRST. In fact, it is necessary to avoid the introduction of extrasystoles or non-dominant beat in the averaging process, since they would alter the quality of the averaged beat. Normally the evaluation of the heart beat type is performed considering its morphology and its occurrence compared to the previous and following beats (rhythm). If the requirement is to obtain a complete rhythm evaluation, then it is necessary an accurate classification of each heart beat based on both morphological and rhythm criteria. However, significant clinical information can be obtained from the analysis of the dominant beat morphology.

For the classification algorithm, only the basic morphological parameters were taken into consideration, trying to limit as much as possible the complexity of such a system. For such purpose, the development and the test of the algorithms were made using the records of the MIT-BIH Arrhythmia Database that includes four records acquired from patients with pacemaker. The algorithm is based on a two-stage clustering technique; firstly a possible classification of all beats is performed, and then all clusters but the one that has been identified with the dominant beats of the signal are reprocessed. In particular, the clusters containing non-dominant beats (according to the first stage) that are large in number are split into smaller ones and reconsidered for misjudgment of being non-dominant. Details will appear elsewhere.

Finally, the averaged dominant beat is represented by the class centroid of the dominant class evaluated on all the QRST assigned to the dominant class after accurate alignment with horizontal and vertical wiggling. Figure 4 shows a graphical interface that, among other functionalities, allows for visualizing the average dominant heart beat and performing linear measurements.

4   Architectural Design and Results

4.1   IT Infrastructure

The signal and image processing methods described in Section 3 have as a result a bunch of clinical parameters together with a new set of annotated images and waveforms (e.g. the segmented echocardiographic sequences and the computed averaged dominant beats). These data should be stored in a structured way in order to trigger CDSS functionalities involving the extracted parameters; further retrieval procedures should be devised to support physicians' case-based reasoning.

Aiming at answering these needs, a composite repository has been prepared and standard-compliant network services have been enabled.

Apart from a standard database for clinical parameters, a DICOM Image Archive has been included into the composite repository. The Image Archive is used to store the original images deriving from a TTE examination as well as the annotated images produced by the image processing toolkit. DICOM Secondary Capture (DICOM-SC) modality is used for the latter purpose, since it is specifically designed to embed the results of image processing (ranging from the application of enhancement filters to more complex image processing procedures) into a DICOM image [14]. The header of the DICOM-SC image may replicate the patient personal information contained in the original DICOM image which is used as input of the image processing algorithms. Further, the header may be used to add a reference to the original DICOM study: in this way the original images and the processed ones are *persistently linked* together within

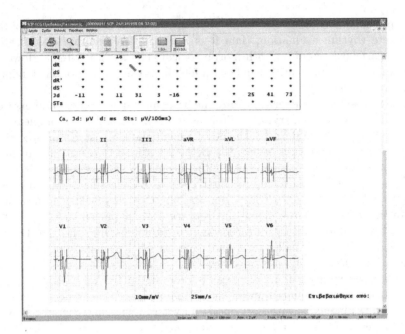

Fig. 4. A screen of the ECG viewer displaying (in zoom mode) additional information including the reference (average) beats. The caliper (ruler) is active and the amplitude and intervals can be accurately measured.

one DICOM study. However, when DICOM-SC is used for storing the results of a segmentation task, a major limitation is represented by the impossibility to edit the segmentation after exporting to DICOM-SC. This problem will be fixed in future releases of DICOM standard; actually some relevant DICOM supplements are in an advanced status of preparation (such as DICOM Supplement 132 which aims at defining the so-called Surface Segmentation Storage SOP Class).

Having obtained in this way an interoperable repository, a second step towards integration consists in embedding network services into the developed prototypical toolkit. Up to now, the image processing toolkit is able to save its results in DICOM-SC format with a meaningful header. The header may replicate the personal details of the patient contained in the original images and other pieces of information which are not altered during processing. A new series UID is associated to the segmented images, while the study UID (if available in the original images) is kept. Further, DICOM utilities (based on the JAVA implementation of DICOM provided by the DCM4CHE toolkit [15]) have been integrated in the toolkit; in particular, the segmented images are sent to the Image Archive directly from the image processing application.

4.2   Integration in the General CDSS Architecture

The intervention of signal and image processing methods into the management of care delivery, as detailed in the previous sections, has been carefully and

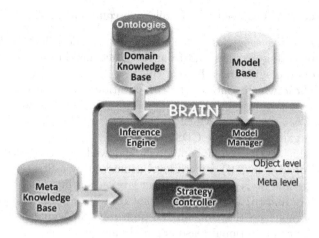

Fig. 5. The CDSS architecture

deeply investigated while designing the CDSS, identifying its functionalities and modeling its architecture. The CDSS has been devised for processing patients' related information by exploiting the relevant medical knowledge which has been opportunely elicited from medical experts and extracted from clinical guidelines. The symbolic paradigm has been selected for formalizing such knowledge into an *ontology-* and *rule-based Knowledge Base* [4]. During the knowledge representation process, the integration of both signal and image processing methods has been conceived in order to embody parameters extracted from different data acquisition modalities into the more general process of health care management. In particular, the integration has been focused on two main issues, i.e. (i) supplying relevant parameters to the inferential processes and (ii) personalizing the diagnostic investigations by suggesting which parameters should be extracted. An example can be used for better explaining the implications of these two issues: while processing a patient's information for identifying the causes of his worsening, the CDSS may need a number of routine parameters not yet available. In such a case, a suggestion will be issued by the system asking the clinician to perform additional examinations, such as an ECG or a TTE, in order to obtain the missing parameters. On the other side, it can happen that such routine parameters are not able to completely explain patient's status and thus the system can require the extraction of other non standard features that can enlighten patient's peculiar conditions. In both cases, the inferential process pauses, waiting for additional information. Reactivating the inferential process requires data processing algorithms to be performed. The CDSS has been hence carefully and specifically designed for incorporating this kind of functioning. Figure 5 shows the CDSS architecture defined according to a multilevel conceptualization strategy which distinguishes between the knowledge and processing components. Such conceptualization division makes the organization of knowledge inside the system explicit, providing an implementation-independent description of the role that various knowledge elements play during the decision supporting process.

The CDSS is then composed by the following components:

- the *Domain Knowledge Base* which maintains the domain knowledge, formalized from the guidelines and from the clinicians' know-how. It consists of a suite of ontologies and a base of rules;
- the *Model Base* which contains the computational decision models, signals and images processing methods and pattern searching procedures;
- the *Meta Knowledge Base* which is composed by the strategy knowledge about the organization of the CDSS tasks;
- the *Brain* which is the system component endowed with the reasoning capability. It is divided into (i) a meta level composed by a Strategy Controller that manages and orchestrates the object level according to what stated into the Meta KB; and (ii) an object level that contained both an Inference Engine for reasoning on the Domain KB and a Model Manager for handling and applying computational reasoning and data processing models.

In particular, the integration of signal and image processing models are, first of all, assured by a dedicated formalization of the relevant acquisition modalities, diagnostic examinations and computable parameters within the ontologies of the Domain KB. Moreover, inferential rules able to process parameters extracted from both signals and images are encoded into the same KB. Finally, the Meta KB contains suitable procedural rules for integrating the application of the data processing methods into the inferential reasoning process. More precisely, when the Inference Engine stops into a crisis status due to the missing values of specific parameters, the Strategy Controller is able to solve the problem by requiring the application of the opportune processing methods triggered by the Model Manager.

5   Conclusions

In this paper we have presented a high-level integration of diagnostic signal and image processing into the wide-ranging services provided by a CDSS for the management of heart failure. In particular, we have motivated the choices made in designing suitably image and signal processing algorithms and we have shown how they can be deployed in decisional problems –and hence in the global process of care– by the CDSS. Future activities will focus on the extension of the already developed signal and image processing toolkit as well as on the realization of an integrated interface for their easy usage in conjunction with the CDSS.

Acknowledgments

This work was partially supported by European Project HEARTFAID "A knowledge based platform of services for supporting medical-clinical management of the heart failure within the elderly population" (IST-2005-027107).

References

1. Doi, K.: Computer-aided diagnosis in medical imaging: Historical review, current status and future potential. Computerized Medical Imaging and Graphics 31, 198–211 (2007)
2. Greenes, D.: Clinical Decision Support: The Road Ahead. Academic Press, London (2007)
3. Swedberg, K., et al.: The Task Force for the diagnosis and treatment of CHF of the European Society of Cardiology, Guidelines for the diagnosis and treatment of Chronic Heart Failure: full text (update 2005). European Heart Journal, 45 pages (2005)
4. Colantonio, S., Martinelli, M., Moroni, D., Salvetti, O., Perticone, F., Sciacqua, A., Chiarugi, F., Conforti, D., Gualtieri, A., Lagani, V.: Decision support and image & signal analysis in heart failure. A comprehensive use case. In: Azevedo, L., Londral, A.R. (eds.) First International Conference on Health Informatics. HEALTHINF 2008. Proceedings, Funchal, Madeira, Portugal, INSTICC - Institute for Systems and Technologies of Information, Control and Communication, pp. 288–295 (2008)
5. Colantonio, S., Martinelli, M., Moroni, D., Salvetti, O., Perticone, F., Sciacqua, A., Gualtieri, A.: An approach to decision support in heart failure. In: Semantic Web Applications and Perspectives (SWAP 2007). Proceedings, Bari, Italy, Dip. di Informatica, Università di Bari (2007)
6. Perner, P.: Introduction to Case-Based Reasoning for Signals and Images. In: Case-Based Reasoning on Signals and Images. Studies in Computational Intelligence, vol. 73, pp. 1–24. Springer, Heidelberg (2007)
7. Perner, P.: Image mining: Issues, framework, a generic tool and its application to medical-image diagnosis. Engineering Applications of Artificial Intelligence 15, 193–203 (2002)
8. Lang, R.M., et al.: Recommendation for Chamber Quantification: A Report from the American Society of Echocardiography's Guidelines and Standards Committee and the Chamber Quantification Writing Group, developed in conjunction with the European Association of Echocardiography, a branch of the European Society of Cardiology. J Am Soc Echocardiogr. 18, 1440–1463 (2005)
9. Noble, J., Boukerroui, D.: Ultrasound image segmentation: a survey. IEEE Trans. Med. Imag. 25, 987–1010 (2006)
10. Barcaro, U., Moroni, D., Salvetti, O.: Left ventricle segmentation in ultrasound sequences for the automatic computation of ejection fraction. In: Open German Russian Workshop on Pattern Recognition & Image Understanding. Proceedings, Ettlingen, Germany, Forshungsinstitut fur Optronik und Mustererkennung (2007)
11. Barcaro, U., Moroni, D., Salvetti, O.: Left ventricle ejection fraction from dynamic ultrasound images. In: Submitted to Pattern Recognition and Image Analysis, Pleiades Publishing House (2007)
12. Chiarugi, F., Sakkalis, V., Emmanouilidou, D., Krontiris, T., Varanini, M., Tollis, I.: Adaptive threshold QRS detector with best channel selection based on a noise rating system. In: Proc. of Computers in Cardiology, vol. 34, pp. 157–160 (2007)
13. Talmon, J.L.: Pattern recognition of the ECG: a structured analysis. PhD thesis, Vrije Universitet of Amsterdam (1983)
14. Zhou, Z., Liu, B.J., Le, A.H.: CAD-PACS integration tool kit based on DICOM secondary capture, structured report and IHE workflow profiles. Computerized Medical Imaging and Graphics 31, 346–352 (2007)
15. DCM4CHE: A Java implementation of the DICOM Standard (2008), http://www.dcm4che.org/

Automatic Data Acquisition and Signal Processing in the Field of Virology

Radu Dobrescu[1] and Loretta Ichim[1, 2]

[1] Politehnica University of Bucharest, Faculty of Automatic Control and Computers,
313 Spl. Independentei, Bucharest, Romania
[2] "Stefan S. Nicolau" Institute of Virology, Romanian Academy
285 Sos. Mihai Bravu, Bucharest, Romania
`radud@isis.pub.ro`, `iloretta@yahoo.com`

Abstract. This paper presents an original experimental optical device (design and construction), based on the Light Scattering Spectroscopy (LSS) and microscopically control of the investigated field. The optical device used for automatic data acquisition in conjunction with a data processing module and fractal analysis. A light beam distributions mathematical model for experimental system is presented. Fractal analysis for Mie light scattering signals are used to extract information about cell nuclei size distributions and could be a useful tool to clearly discriminate between non-infected and virus infected cultures.

Keywords: Mie scattering, optical device, signal analysis, cell nuclei, fractal analysis.

1 Introduction

Until 1998, quantitative evaluation for transformation degree could be done using optical invasive methods: morphometry, microscopy and flow cytometry. Histology and morphometry require tissue removal, processing, and microscopic examination to reveal morphological characteristics. In flow cytometry, cells suspended in a fluid are ejected from nozzle and made to flow, and are then sorted using laser light according to size and shape, using angular light scattering properties or fluorescence from attached flurophores.

Recently, Perelman and collaborators [1] proposed a new spectroscopic method has been developing using the Mie theory, started from the necessity of cellular structures analysis, with a view to establish the amount of cellular nuclei size distribution. Changes in nuclei of the epithelial cells are amongst the most important indicators for dysplasia. Some of the major diagnostic criteria include nuclear enlargement, increased variation in nuclear size and shape (pleomorphism).

Mie scattering could be used to obtain a cell nuclei size distribution and correlate the findings with pathological aspects of the sample [2-4] from backward Light Scattering Spectroscopy (LSS) data in the visible spectral range.

One important application of the technique capable of measuring quantitative changes of subcellular shape and structure *in situ* is early diagnosis of cancer or precancerous lesions [2]. The LSS is a non-invasive optical technique capable of

P. Perner and O. Salvetti (Eds.): MDA 2008, LNAI 5108, pp. 52–61, 2008.

providing quantitative information that need further processing in order to be used in discrimination between different types of cells [5]. The light beam entering the cell holder surface and the outgoing beam are brought close together in a fibre optic attachment mounted below the microscope sample holder.

The aim of the present research was to design an improved method [1] for measuring, under microscope control of the investigated field, the backward Mie LSS energies crossing cell cultures on transparent surfaces of various cell holders, in order to correlate the fractal dimensions (Fd) for cell nuclei size distribution changes with pathological processes.

2 Materials and Methods

2.1 Instrumentation

The optical device represents a system which allows the recording of the backward diffusion spectrum by recording the light reflectance gathered from a sample mounted onto the microscope stage. The recorded spectrum is analyzed *on line* using software tools with a view to obtain data about cellular nuclei size distribution observed in the optical field of the object lens.

The optical device was patented [6] and the main components are: light source, diode array multichannel UV-VIS spectrophotometer, optical fibre guides, optical connectors, lens and neutral filters.

The schematic diagram of the device is presented in figure 1.

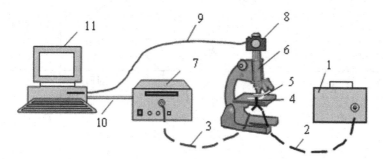

Fig. 1. Block diagram of the optical device: 1 – light source; 2, 3 – optical fibre guides; 4 – diffusiometric probe; 5 – sample; 6 – microscope; 7 – spectrophotometer; 8 – digital camera; 9 – USB cable; 10 – computer spectrophotometer interface; 11 – PC.

In particular case, the optical device includes:

- light source comprises collimated light source with a NARVA 6V–20W halogen lamp, transformer (6V – 30W), biconvex lens, optical neutral filters;
- two optical fibre guides quartz Fibroflex with terminal parts CZ No. 772610-9018 (Zeiss – Jena), length $L = 200$ cm and diameter $d = 1$ mm;
- two optical connectors (Zeiss – Jena) length $L = 2$ cm, diameter $d = 2$ mm;
- optical microscope ML – 4M IOR;

- digital Creative PC-CAM 300 camera;
- diffusiometric probe 2 x $d = 1$ mm / 1 x $d = 1.42$ mm (Zeiss – Jena);
- MCS 420 Zeiss diode array multichannel UV-VIS spectrophotometer;
- Pentium III computer with interface for MCS 420 Zeiss multichannel spectrophotometer.

2.2 Light Beam Distributions Mathematical Model

To highlight all parameters used in the experimental system it was proposed a light beam distribution mathematical model, which cross analyzed samples.

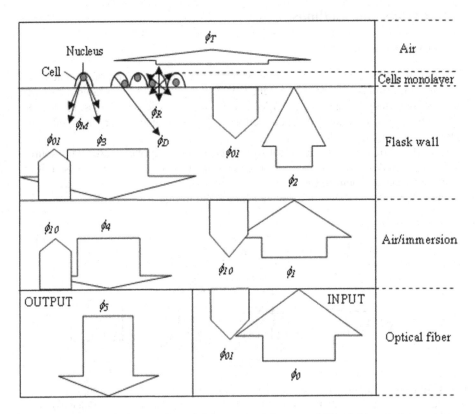

Fig. 2. The main optical parameters of the system used to record Mie light scattering spectrum crossing cell cultures samples

Using the optical device presented above they were automatically recorded Mie light scattering spectra corresponding to analyzed cell cultures. Automatically acquisition for these spectra occurs in visible range using length between $\lambda = 350 - 619$ nm.

The main optical parameters of the system are described in figure 2 where Mie LSS energy outputs recorded from a cells culture monolayer in terms of the input light energy are used in a theoretical approach of the optical phenomena involved.

The method is based on light energy output ϕ_e measurements of a sample versus those of a reference $\phi^o{}_e$. In these conditions, difference spectra were computed using the following equation: $(\Delta\phi_e) = (\phi_e) - (\phi^o{}_e)$.

The peculiarities of the spectra are expressed in terms of all parameters involved as it is presented below.

The light energy flow coming from the light source through the first optical fibre guide and reaching the tip front is ϕ_0; a fraction of the energy is reflected backwards from the quartz fibre in the air with a ρ_{01} coefficient so that the emergent energy fraction ϕ_1 becomes:

$$\phi_1 = \phi_0 - \phi_{01} = \phi_0(1 - \rho_{01}).$$ (1)

The beam energy flow ϕ_2 reaching the flask is reflected at the air-flask interface with a ρ_{10} coefficient, resulting in:

$$\phi_2 = \phi_1(1 - \rho_{10}) = \phi_0(1 - \rho_{01})(1 - \rho_{10}).$$ (2)

The light energy flow ϕ_3 reaching the lower surface of the flask represents the algebraic sum of all three components ϕ_M (represents the Mie light scattering by the nuclei of the N cells occupying an Ns surface fraction of the beam), ϕ_R (represents the Rayleigh scattering component of the cell biopolymers not collected by the signal output optical guide and ϕ_D (is the sum of diffuse reflections due to the rough surface of the cell membranes in contact with the air):

$$\phi_3 = \phi_M + \phi_R + \phi_D = \phi_0(1 - \rho_{01})(1 - \rho_{10})\{[1 + Ns(k - 1)]\rho_{01} + Ns(\mu - \delta)\},$$ (3)

The next reflection at the flask-air interface diminishes by reflection with a $(1 - \rho_{01})$ coefficient the ϕ_3 outgoing beam, so that ϕ_4 becomes:

$$\phi_4 = \phi_3(1 - \rho_{01}) = \phi_0(1 - \rho_{01})^2(1 - \rho_{10})\{[1 + Ns(k - 1)]\rho_{01} + Ns(\mu + \delta)\}.$$ (4)

In an analogous way, ϕ_5 – the light transmitted through the fibre guide to the MCS 420 spectrophotometer - becomes:

$$\phi_5 = \phi_0[(1 - \rho_{01})(1 - \rho_{10})]^2\{\rho_{01} + Ns[(\mu + \delta) + (k - 1)\rho_{01}]\}.$$ (5)

The measured light energy ϕ_e needs also a correction factor C, due to other losses along the quartz fibres and the reflections at the end adapted at the spectrophotometer entrance. The corrected equation is then:

$$\phi_e = \phi_0\{[(1 - \rho_{01})(1 - \rho_{10})]^2\{\rho_{01} + Ns[(\mu - \delta) + (k - 1)\rho_{01}]\} + C\}.$$ (6)

For the cell free reference, when $Ns = 0$, the above equation becomes:

$$\phi^o{}_e = \phi_0\{\rho_{01}[(1 - \rho_{01})(1 - \rho_{10})]^2 + C'\}$$ (7)

and the difference spectrum $\Delta\phi_e = \phi_e - \phi^o{}_e$ of the sample in terms of the reference and the term $\phi_0\Delta C = C - C'$ represents an apparatus constant, easy to eliminate from the energy values recorded for each spectrum at 350 nm, the equation becomes:

$$\Delta\phi_e = Ns\phi_0[(1-\rho_{01})(1-\rho_{10})]^2\{\mu-[(1-k)\rho_{01}+\delta]\}. \tag{8}$$

There are at least three comments to be made on the result obtained: 1) it is obvious from the last equation that the scattering is proportional to the area fraction Ns of the measuring beam section occupied by all cells and become nil for cell free samples; 2) Mie scattering patterns are recorded as positive peaks of the $[(1-k)\rho_{01}+\delta]$ diffuse reflection and Rayleigh scattering spectra, which are diminishing continuously the energy levels of the recorded spectrum; 3) the $\phi_0\Delta C$ correction term could be avoided using technical methods during processing phase.

When measurements were performed with a transparent immersion liquid, e.g. silicon oil, between the light guide tip and the cell holder so that $\phi_0 = \phi_1 = \phi_2$, and $\phi_3 = \phi_4 = \phi_5$, corresponding equations are resulting in:

$$(\phi_e)_{im} = \phi_0\{[\rho_{01} + Ns\{\mu-[(1-k)\rho_{01}+\delta]\}]+C\}, \tag{9}$$

and respectively, and the corresponding value for the reference:

$$(\phi^o_e)_{im} = \phi_0(\rho_{01}+C'), \tag{10}$$

so that, after corrections, the finally measured light energy difference becomes:

$$(\Delta\phi_e)_{im} = (\phi_e)_{im} - (\phi^o_e)_{im} = Ns\phi_0\{\mu-[(1-k)\rho_{01}+\delta]\}. \tag{11}$$

By comparing equations (8) and (11) it is obvious that $\Delta\phi_e$ is decreased relative to $(\Delta\phi_e)_{im}$ by a coefficient $[(1-\rho_{01})(1-\rho_{10})]^2 <1$.

It must be emphasized that there must also be another distinction between the results obtained with and without immersion liquid, shown by a shift of the Mie scattering peaks in the difference spectra calculated. That is due to the fact that this backward scattered light is not perpendicular to the flask surface, but specifically disposed in axial symmetric cones corresponding to different solid angles. The backward coming photons reaching the output light guide are more deviated at the two interfaces separating the flask from the fibre guide tip when air is between them, than when refraction indexes are equalized by an immersion liquid.

From the experimental data obtained on hand of both measurement methods the value of $[(1 - \rho_{01})(1 - \rho_{10})]^2$ could be appraised, expressing the overall signal to noise ratio increase when no immersion liquid between the light guide tip and the sample is used.

$$\Delta\phi_e/(\Delta\phi_e)_{im} = [(1-\rho_{01})(1-\rho_{10})]^2 \tag{12}$$

Considering that the reflection coefficients ρ_{01} and ρ_{10} are, in a first approximation, equal ρ, their value could be appraised from the fourth root of the $\Delta\phi_e/(\Delta\phi_e)_{im}$ ratio:

$$\rho = [\Delta\phi_e/(\Delta\phi_e)_{im}]^{1/4}. \tag{13}$$

2.3 Cell Cultures

To estimate changes in cell nuclei size distribution using fractal analysis method, an experimental model for *in vitro* infection with Herpes Simplex Virus (HSV) in human cervix epithelial carcinoma cell line was implemented (HeLa CLL No. 2). It was using LOV strain for type two HSV. This line is used currently in virology studies owing to its susceptibility in HSV virus infection. The reference used was represented by non-infected HeLa culture. The culture conditions have been described in [7].

2.4 Automatic Data Acquisition and Processing

In figure 3, it was presented the data acquisition and processing diagram. This consists of two modules.

The left hand module presents the optical device, which recorded the Mie scattering spectra. Here, it was presented analysis of complexity of spectra collected with the optical device. The outgoing beam spectrum was recorded with a personal computer system Intel, Pentium III assisted diode array MCS 420 (C. Zeiss – Jena) simultaneous spectrophotometer, and the corresponding commercially available ASPECT software.

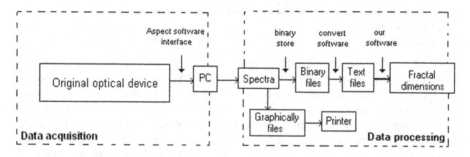

Fig. 3. Data acquisition and processing diagram

On the right hand, it was represented the data processing module, which accomplishes three functions:

1. process spectra to compute their fractal dimensions, using our own software package;
2. store spectra to create a file database, which allow permanent access to any already processed spectrum;
3. print or view any processed spectrum.

To verify that only in the presence of cells, the Mie LSS spectra recorded in the visible domain show the characteristic pattern of several equidistant peaks [1], a non-infected HeLa cell monolayer on a culture flask and a HSV-infected HeLa cell monolayer on a culture flask were used as samples, and a cell free part of the flask tissue as reference. The cells were cultivated in 25 cm^2 flask (Falcon, Becton Dickinson). As immersion liquid between the optical fibre attachment tip and the cell holder wall, Merck Silicon oil 7742 was used. The method is based on light energy output measurements of a sample vs. those of a reference.

For non-infected HeLa cells, it was registered spectra from ten different areas from culture flask, microscopic chosen. But, for HSV-infected HeLa cells, the culture flask was scanned each 5 mm length and each 4 mm width, so spectra from the whole flask area was collected. In this way, they were automatically obtained spectra for 60 different areas from the flask's HSV-infected HeLa cells because of a high level of heterogeneity owing to cellular nuclei diameter variation, function of infection degree.

Ten individual spectral curves were recorded for each reference and ten individual spectral curves were recorded for each sample, for each investigated area. Mie scattering spectra were registered in the same experimental conditions.

In many applications based on spectral tools, data are registered using a conventional way and then processed using a transformation algorithm. So, Mie scattering spectra registered using optical device needed extra processing to remove unexpected effects (noises, perturbations).

An optimized spectral data processing algorithm consists of the following steps:

1. Compute averaged Mie diffusion spectra;
2. Bring all ordinates at the origin and make needed corrections;
3. Smooth spectra;
4. Subtract sample spectra from the reference ones;
5. Normalize spectra;
6. Subtract normalized spectra from mathematically simulated smoothed spectra.

Signal complexity can be analyzed either directly in time domain, or in frequency domain. Analysis in the frequency domain requires Fourier or wavelet transform of the signal. But fractal analysis may be done directly in time domain. In time domain calculation of fractal dimension, is much quicker and so easier to be done in real time. Regarded signal analyses, a variety of methods are available to compute the fractal dimension, Fd representing the measure of the shape irregularity. To determine the fractal dimension, below denote Fd value, data were treated using Hurst method [8]. We have to stress that fractal dimension in this case is always between 1 and 2, since it characterizes complexity of the curve representing the signal under consideration on a two-dimensional plane.

3 Experimental Results

Figure 4 shows two Mie light scattering spectra examples in visible range, after completing the processing algorithm: (a) investigated aria 1 spectrum of non-infected HeLa cells flask and (b) investigated aria 1 spectrum of HSV-infected HeLa cells flask.

Mie light scattering spectra data obtained from HSV-infected and non-infected HeLa cells were stored in separate files. Using fractal dimension algorithm and corresponded software package, all spectra data files for each area involved were processed. The algorithm used to compute the fractal dimension have been proposed and applied to experimental data derived from cells light scattering spectrum. Fractal dimension values were calculated in the spectral range between $\lambda = 400$ and 600 nm, using a 201-points window. The algorithm was implemented in Microsoft Visual C++ 6.0 software.

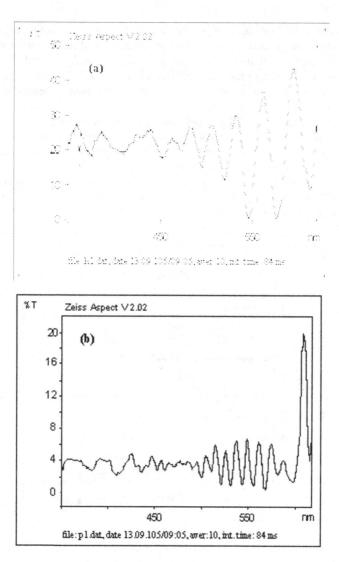

Fig. 4. Processed Mie light scattering spectra: a) aria 1 spectrum of non-infected HeLa cells flask; b) aria 1 spectrum of infected HeLa cells flask

Mie light scattering spectra data obtained from HSV-infected and non-infected HeLa cells were stored in separate files. Using fractal dimension algorithm and corresponded software package, all spectra data files for each area involved were processed. The algorithm used to compute the fractal dimension have been proposed and applied to experimental data derived from cells light scattering spectrum. Fractal dimension values were calculated in the spectral range between $\lambda = 400$ and 600 nm, using a 201-points window. The algorithm was implemented in Microsoft Visual C++ 6.0 software.

Fractal dimension, the aim of signal analysis data processing program, is a parameter, which characterizes investigated cell nuclei size distribution complexity.

Based on experimental data were computed average and standard deviation for fractal dimensions related to non-infected HeLa cellular nuclei size distribution, for a group of ten investigated areas using Hurst method. The result computed using Hurst method is: Fd = 1.357 ± 0.02.

So, fractal dimensions for HSV-infected HeLa cellular nuclei size distribution computing using Hurst method needed first of all an ascendant average for values, followed by five value groups dividing. After that they were computing the average and standard deviation for each group. It was obtained 12 groups.

Based on computed averages for fractal dimension groups, a graphic for non-infected and HSV-infected cells was plotted using Hurst method (fig. 5) to identify the distribution way of obtained values.

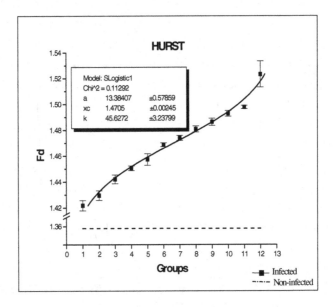

Fig. 5. Representation by groups for fractal analysis averages computed using Hurst method

It comes out that fractal dimension averages for HSV-infected cells have a type of the following equation:

$$y = \frac{a}{1 + e^{-k(x - x_c)}}, \tag{14}$$

where a, x_c and k are real parameters.

Using software product MicroCalc Origin 5.0, there was represented a graphic for fractal dimension averages distributions by groups for above equation. Parameters of the logistic equation are: a = 13.384 ± 0.578; x_c = 1.47 ± 0.002; k = 45.627 ± 3.238 and χ^2 = 0.113.

In Hurst method case comes out that all non-infected HeLa cells have fractal dimensions average equal to 1.357. All HeLa cells groups having fractal dimensions averages higher then 1.422 were HSV-infected. Value 1.422 represent a threshold in Hurst method, which separates virus infected and non-infected cells.

4 Conclusions

The created optical device allowed to evidence cellular changes for *in vitro* cultures analyzed using Mie spectra. Resulted spectra are according to Mie theory.

It was founded that the fractal dimension of cell nuclei size distribution increased as the degree of HSV-infected HeLa cells increased. It comes out that fractal dimensions values present much higher variations in case of infected cultures, which is contrary to non-infected ones that have constant, reduced size nuclei.

The study has demonstrated a useful, practical tool based on fractal dimension to discriminate between virus infected and non-infected biological samples.

References

1. Perelman, L.T., Backman, V., Wallace, M.B., Zonios, G., Manoharan, R., Nusrat, A., Shields, S., Seiler, M., Lima, C., Hamano, T., Itzkan, I., Van Dam, J., Crawford, J.M., Feld, M.S.: Observation of periodic fine structure in reflectance from biological tissue: A new technique for measuring nuclear size distribution. Phis. Rev. Lett., 627–630 (1998)
2. Backman, V., Gurjar, R., Badizadegan, K., Itzkan, I., Dasari, R.R., Perelman, L.T., Feld, M.S.: Polarized light scattering spectroscopy for quantitative measurement of epithelial cellular structure in situ. IEEE J. Quant. Electr. 5, 1019–1026 (1999)
3. Sokolov, K., Drezek, R., Gossage, K., Richards-Kortum, R.: Reflectance spectroscopy with polarized light: is it sensitive to cellular and nuclear morphology. Optics Express 5, 302–317 (1999)
4. Myakov, A., Nieman, L.T., Wicky, L., Savchenko, N., Utzinger, U., Richards-Kortum, R.: Fiber optic probe for polarized reflectance spectroscopy in vivo: design and performance. J. Biol. Opt. 7, 388–397 (2002)
5. Mourant, J.R., Johnson, T.M., Carpenter, S., Guerra, A., Aida, T., Freyer, J.P.: Polarized angular dependent spectroscopy of epithelial cells and epithelial cell nuclei to determine the size scale of scattering structures. J. Biol. Opt. 7, 378–387 (2002)
6. Ichim, L.: Spectrophotometric device for microscope. RO patent no. 120590 (2006)
7. Mutiu, A., Alexiu, I., Chivu, M., Petica, M., Anton, G., Bleotu, C., Diaconu, C., Popescu, C., Jucu, V., Cernescu, C.: Detection of human papillomavirus gene sequences in cell lines derived from laryngeal tumors. J. Cell. Mol. Med. 49 (2001)
8. Hurst, H.E., Black, R., Sinaika, Y.M.: Long-term storage capacity of reservoirs: An experimental study. Constable, London (1965)

Colorectal Polyps Detection Using Texture Features and Support Vector Machine

Da-Chuan Cheng[1], Wen-Chien Ting[2], Yung-Fu Chen[3], Qin Pu[4], and Xiaoyi Jiang[5]

[1] Department of Radiological Technology, China Medical University, Xueshi Road 91, Taichung, Taiwan
dccheng@mail.cmu.edu.tw
[2] Colorectal Surgery, China Medical University Hospital, Yude Road 2, 404, Taichung, Taiwan
milka3670@pchome.com.tw
[3] Department of Health Services Administration, China Medical University, Xueshi Road 91, Taichung, Taiwan
yungfu@mail.cmu.edu.tw
[4] Department of Internal Medicine, University Hospital of Freiburg, Freiburg, Germany
pujinshuo@yahoo.com
[5] Department of Mathematics and Computer Science, University of Münster, Einsteinstr. 62, 48161, Münster, Germany
xjiang@math.uni-muenster.de

Abstract. In this paper we propose a novel method in detecting colorectal polyps on colonoscopic images. Texture features are applied in polyps and normal tissues training and classification. Support vector machine is used as a classifier to identify the position of polyps. Seventy-four colonoscopic images are collected to test the system. Half of them are used as training images and half are used as testing. The experimental result shows the system can identify all polyps if the colonoscopic images contain single polyp. The sensitivity is 86.2% and the false-positive rate is 1.26 mark per-image.

1 Introduction

Colorectal cancer is the third leading cause of caner-related deaths in the United States in 2003 [1]. In this report, an estimated 105,500 colon and 42,000 rectal cancer cases are expected to occur and about 57,100 deaths are expected to occur in 2003. The death due to colon cancer can be prevented and cured through early detection. Therefore, early diagnosis is critically important for patient's survival.

Nowadays, two kinds of screening are common for early polyps detection: CT and colonoscopy. The examination using CT has certain advantages. However, the disadvantages are:

1. once some polyps are detected, the patient has to be re-examined via colonoscopy to remove the polyps.
2. some small and tiny polyps are difficult to be detected. If such polyps remain in the colon, they can possibly grow into malignant lesions.

Colonoscopy is an accurate screening technique for detecting polyps of all size, which also allows for biopsy of lesion and removal of most polyps [2].

P. Perner and O. Salvetti (Eds.): MDA 2008, LNAI 5108, pp. 62–72, 2008.

The eventual goal of our study is to develop an automatic system which is able to detect the colonic polyps on colonoscopic images in real-time. This system would be available in helping physicians to notice some polyps which are of small or tiny size. Normally, polyps of large and middle size can be easily found by physicians. However, during the surgery the physicians have less time to notice those polyps of small or tiny size. Therefore, a CAD (Computer Aided Detection) system would be useful in polyps detection. To achieve this goal, a fast polyps detection algorithm should be developed and connected to the video capture system.

The goal of this paper is to explore a method which is able to detect polyps in static images. In this stage, the speed of polyps detection is not the key point. Instead, the accuracy of detecting polyps is our intended achievement.

The previous related studies on polyps detections can be categorized into two kinds: 1) CT based and 2) colonoscopy based method. Generally, the CT-based method [3,4] uses morphology such as the curvature (2D or 3D) as feature to recognize polyps. This is because in CT images there are no texture information. The colonoscopy-based method uses in general two kinds of features: 1) shape (or morphology) and 2) texture information. In general, using texture is an advantage in analyzing colonoscopic images. The reason is [5]: The colonic mucosal surface is granular and demarcated into small areas. Changes in the cellular pattern (pit pattern) of the colon lining might be the earliest sign of polyps [6]. These texture alterations of the colonic mucosa surface can also be used for the automatic detection of colorectal lesions.

Texture features are very common in computer vision to classify textures or recognize different object via objects' surface texture analysis. Haralick [7] and Cohen [8] have proposed some texture features which are very popular to image processing, pattern recognition, and computer vision. In particular, co-occurrence matrices approach is very common and useful in texture feature extraction, texture analysis, and texture classification.

The most related study to ours is using color wavelet features in detecting colorectal polyps [5]. They used wavelet transform because of its advantage on multi-resolution analysis. This multi-resolution analysis is on spatial as well as on frequency domain. Two filters were used to extract the information: low-pass and band-pass filters. Both filters were applied to all levels of different resolutions. Afterwards, the co-occurrence matrices approach was used to extract texture features. Four features were derived from the co-occurrence matrices: angular second moment, correlation, inverse difference moment, entropy, contrast [9], and dissimilarity [10]. The classification is performed via linear discriminant analysis (LDA). They claimed that using more complex classifier would increase the number of parameters associated with the evaluation of the proposed feature set.

Another similar work [11] used texture features as well and compared many tools of analyzing features. These tools included Backpropagation Neural Network (BPNN), resilient propagation (RPROP), scaled conjugate gradient (SCG), and Marquardt algorithms. They combined PCA and BPNN together as a classification tool in discriminating normal and abnormal colon status.

Another study is similar to colorectal polyps analysis but on gastric polyps in endoscopic video [12]. In this study, they compared four texture features to find their performance in gastric polyps detection. The features they have compared are texture spectrum histogram, texture spectrum and color histogram statistics, local binary pattern histogram, and color wavelet covariance (CWC). Their finding was that the CWC feature is the best feature among all which reaches 88.6±2.3%.

The proposed detection scheme involves a) a novel feature extraction technique based on co-occurrence matrices with a simple difference of texture features of different color channels. b) the support vector machine is applied as a classification tool in our polyps detection scheme.

The rest of the paper is organized as follows. In Section 2 we describe the method and equipments how the colorectal images are acquired. The texture features are proposed in Section 3. In Section 4 the training and classification processes are issued. The experimental results are shown in Section 5 and the conclusion is given afterwards.

2   Colonoscopic Image Acquisition

The endoscopic video system (Fujinon EPX-402 & EC-410) is used for whole colon examination. Before examinations, patients are asked to receive colon preparation in 2 days with laxative in order to clean the residua. The colonoscopy is intruded into the cecum via anus and air is inflated via colonoscope tube into colon lumen to achieve a full view of the colon mucosa. During examination videos are shown on the monitor and are recorded. The polyp images are digitally captured into a computerized reporting system and are saved in BMP or JPG format.

The image's spatial resolution is 378×254 pixels in three colors (R, G, and B). We have collected a total of seventy-four colonoscopic images to test our system. Among them thirty-seven images are used as training images and thirty-seven images are used as testing.

3   Textural Feature Extraction

3.1   Co-occurrence Matrix

The gray level co-occurrence matrix (GLCM) is a well-known and popular technique for extracting texture information from images. Conner and Harlow [13] have shown that GLCM is a more powerful technique than gray level difference matrix (GLDM), gray level run length method (GLRLM), and the power spectral method (PSM). Ohanian and Dubes [14] showed results that GLCM features perform better than fractal, Markov Random Field, and Gabor filter features for classification of texture.

Over twenty features appear in the literature which can be used to extract information from co-occurrence matrices [15]. Most of them work on gray level image. The color co-occurrence matrix (CCM) is seldom applied because of its lower efficiency in computation time. However, colonoscopic images are color images where color is a very important feature in polyps detection tasks. Omitting the color information will cause difficulties in discriminating colonic polyps and normal tissues. In this paper, we propose a method which can apply color texture information.

3.2 Color Texture Feature

Our colonoscopic images have three color channels: R,G, and B. Based on the calculation, it is found that the green and blue channel images are good in discriminating
normal and abnormal mucosal surface ($p < 0.0001$ using F_1 on each channel image).
On the contrary, the red channel image has some strong light reflection ($p = 0.0310$).
Moreover, the cross-correlation coefficient between green and blue channel is larger
than 0.8. Therefore, the smoothness feature can be applied either on green or blue channel. Here we chosed green channel. Let I denote a colonoscopic image and let I_s denote
an arbitrary sub-image sampled from I. All features (F_1 - F_8) as follows are calculated
based on the sub-image I_s:

$$F_1 = 1 - \frac{1}{1 + \sigma_g}$$

where σ_g denotes the standard deviation of the green channel image.

It is found that the colonoscopic images have different colors. The average intensity is then a useful feature. The second texture feature is simply a combination of the
average intensity of all three channel images:

$$F_2 = \frac{\bar{r}^2}{\bar{g}\bar{b}}$$

where $\bar{r}$, $\bar{g}$ and $\bar{b}$ are the mean intensity of the red, green, and blue channel images,
respectively. However, the ratio of the mean intensity cannot represent the difference
very well. Two measurements are added as features:

$$F_3 = \frac{\bar{r} - \bar{g}}{256}, \quad \text{and} \quad F_4 = \frac{\bar{r} - \bar{b}}{256}$$

Let the co-occurrence matrix (CM) of a certain sub-image I_s with an offset $(0, 1)$ be
denoted as $p(i, j)$. The CM made from green and red channel images are represented as
$p_g(i, j)$ and $p_r(i, j)$, respectively. The contrast and energy features are utilized. However, a small modification is proposed as follows.

$$F_5 = \sum_{i,j} |i - j|^2 p_g(i, j) - \sum_{i,j} |i - j|^2 p_r(i, j)$$

$$F_6 = \sum_{i,j} p_g(i, j)^2 - \sum_{i,j} p_r(i, j)^2$$

The difference of contrast is used as a feature instead of the original one.

Another co-occurrence matrix with different offset $(-2, 0)$ is utilized and denoted as
$q(i, j)$. This matrix is made from I_s as well. Similar features are defined:

$$F_7 = \sum_{i,j} |i - j|^2 q_g(i, j) - \sum_{i,j} |i - j|^2 q_r(i, j)$$

$$F_8 = \sum_{i,j} q_g(i, j)^2 - \sum_{i,j} q_r(i, j)^2$$

All features $F_1, F_2, \cdots, F_8$ form a feature vector to describe a certain sub-image I_s.

3.3   Feature Normalization

Let N_1 and N_2 denote the number of sub-images of normal and abnormal tissues in the database for training. Each sub-image has a feature vector $v = [F_1 F_2 \cdots F_8] \in \mathbb{R}^{8 \times 1}$. All feature vectors form a feature matrix $[v_{n,1} \; v_{n,2} \; \cdots \; v_{n,N_1} \; v_{a,1} \; v_{a,2} \; \cdots \; v_{a,N_2}] \in \mathbb{R}^{8 \times (N_1 + N_2)}$, where v_n and v_a denote the feature vector of normal and abnormal sub-image, respectively.

It is very common to normalize features before using them in computer vision. In this study, we simply choose the maximum value in the matrix and the feature matrix is divided by this maximum to keep all elements less or equal to one. This normalized feature matrix is used to train the classifier stated in section 4.2.

4   Polyps Training and Classification

4.1   Training Patterns

The training patterns contain polyps and normal tissue sub-images. All sub-images are selected via a GUI (Graphic User Interface) on Matlab platform [16] by a well-trained expert. The size of all sub-images are the same in 33×33 pixels with R,G, and B channels. The size is determined due to following three reasons: 1) There are different sizes of polyps in our colonoscopic images. Some polyps are very large and some are smaller than 33×33. If the window size is too large, it is possible to include the normal tissue in the window and the abnormal to normal ratio might be less than 50%. Under this situation, the feature of polyps might cause ambiguities in discrimination. On the contrary, if the window is too small, it is hard to represent the texture changes in a certain area. The larger the population of pixels in the sub-image, the more informative the features are expected to be. 2) In [12], they chosed 32×32 as a window size. In their study, they selected the abnormal tissues with exclusion of the region with strong light reflections. Based on this criterion, the window size cannot be too large. In our images, we have also strong light reflections on polyps. Although we have no such criterion, it is not good to include a large area of strong light reflections. Under this consideration, the window size cannot be too large. 3) Partial inclusion of strong light reflections in the sub-image is allowed in our scheme because the reflection is also a good feature in representing a polyp. It is found that most polyps contain more or less strong light reflections.

Our database contains sub-images with normal tissues and abnormal tissues (polyps). The pattern number of normal tissues is larger than the abnormal ones. This is because most images contain only one or two polyps and most regions are normal tissues in an image. Therefore, it is necessary to select more sub-images as training patterns for normal tissues.

Some criteria of selecting training patterns are considered. The selected sub-images from a polyp image are in general not overlapping. This criterion works also in selecting polyp sub-images. Figure 1 shows one typical colonoscopic image.

The selection of polyps' sub-images is in general not overlapping. The strong light reflection is considered as a feature so that it can be included in sub-images. However, the region of the light reflection should not be too large. It is suggested to separate

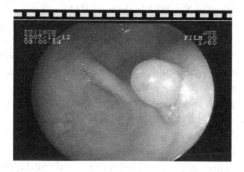

Fig. 1. A typical colonoscopic image with a polyp. There are some strong light reflections on the mucosal surface.

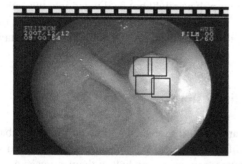

Fig. 2. Polyps sub-image selection is in general not overlapping. The strong light reflection cannot occupy too large region in a sub-image.

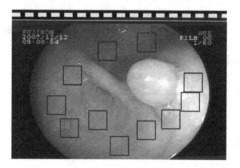

Fig. 3. The number of sub-images of normal tissue is in tendency more than the abnormal tissues

them into different sub-images as shown in Fig. 2 Moreover, if the polyp is large, it is suggested to select as much sub-images as possible to cover the whole polyp.

The selection of sub-images of normal tissues included the dark regions, the reflections and so on. We suggest selecting partial reflections in a sub-image as well. This is because there are also reflection on normal mucosal surfaces; see Fig. 3.

4.2 Classification: Support Vector Machine (SVM)

Support Vector Machine (SVM) [17] is a kernel-based classifier. It is very popular in pattern recognition and computer vision. Many applications are based on SVM to have a nonlinear classification. In this study, we utilize this tool to be a two-class classifier. The radial basis function is selected as the kernel function.

A rectangle active region is defined in every test image. This rectangle area is of the same size and the same position in each image and it includes most colonic mucosal surfaces in images. This definition reduces the redundant calculation on the image border without useful information.

Every test image is automatically divided into several overlapping sub-images. The sub-image is of the same size to the one in the training process, i.e., 33×33 pixels with three channels (R, G, and B). The occlusion ratio is 50% in both x- and y-direction, respectively, i.e. the sliding window has 17 pixels translation in both directions. Under this construction, smaller polyps might not be divided into two sub-images.

All sub-images are fed into our well-trained SVM-classifier to test if it contains polyps or not. If one sub-image is labeled as a polyp image, a rectangle will be marked on that position for observation.

5   Results and Discussions

A total of 74 images are carefully selected and saved in our database. They are representative containing different types of polyps, different colors and so on. Half of them are used as training patterns (images) and half of them are used as test images.

Sixty-one sub-images are manually selected by an expert as polyps sub-images and 283 sub-images are selected as normal tissues from the 37 training images. Figure 4 are some examples of training sub-images of polyps. There are less or more strong light reflection spots on polyps surfaces. Figure 5 shows some example sub-images of normal tissue in the training patterns. Most polyps sub-images have strong light reflection spots. Some sub-images of normal tissue have vessels and this is a good feature to be distinguished from polyps.

The normalized cross-correlation coefficient is a measurement if two signals are similar to each other or not. If the value between two signals is small then it is less

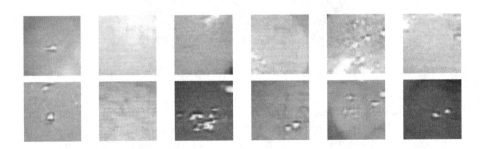

Fig. 4. Twelve from sixty-one polyps sub-images in the training patterns

Fig. 5. Twelve from 283 sub-images of normal tissue in the training patterns

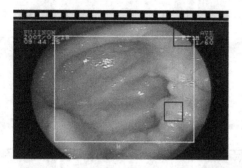

Fig. 6. The multiple polyps in the image are not well detected

Table 1. The normalized cross-correlation coefficients of all features

1	2	3	4	5	6	7	8	9	
1.0	-0.2441	-0.0490	0.1008	0.1890	0.3410	-0.1878	0.3534	-0.1562	1
	1.0	0.4542	0.5738	0.4368	-0.1034	0.3621	-0.0976	0.3258	2
		1.0	0.7374	0.1987	0.0378	0.0061	0.0270	-0.0164	3
			1.0	0.8085	0.1556	0.0908	0.1166	0.0630	4
				1.0	0.1929	0.1264	0.1457	0.1056	5
					1.0	-0.1298	0.8676	-0.0741	6
						1.0	-0.0822	0.9819	7
							1.0	-0.0240	8
								1.0	9

correlated. This can be treated as a tool to see if two features are correlated to each other. The results are shown in Table 1.

Thirty-seven images are tested. Some contain one single polyp in an image and some contain multiple polyps. All polyps in single-polyp images are detected. However, some small polyps are not detected in the multiple-polyps images. One example is shown in Fig. 6. This might be due to that we did not choose tiny polyps in the training patterns. The sensitivity is 86.2% and the the false-positive rate per-image is 1.26.

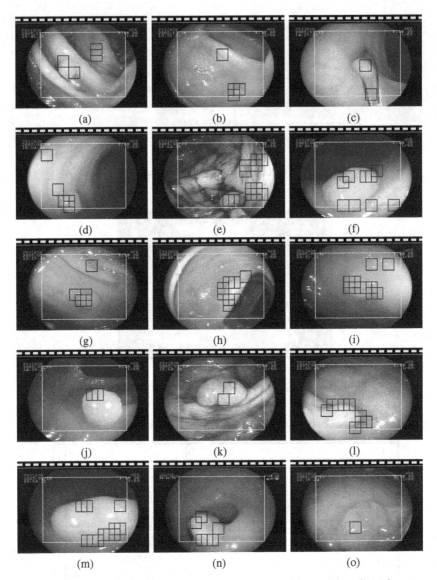

Fig. 7. Polyps detection results. (a)-(i) have correct marks and false-positive. (j)-(o) have correct marks without false-positive.

Figure 7 shows some detection results, in which (a)-(i) have correct marks and false-positive and (j)-(o) have correct marks without false-positives. Two rectangles have correctly marked the polyp and one false-positive in Fig. 7(a). This might due to that the color is similar to polyp's color. Although two rectangles marked as false-positive, we calculate only one false-positive since they are overlapping.

Figure 7(b) shows another detection result. A small polyp (right-bottom) is detected. Figure 7(c-d) and (f) detect the polyps. The false-positive might be because of the smooth mucosal surface which is like a polyp. In Fig. 7(f) and (i), the false-positive have small strong light reflection spots and the mucosal surface is smooth. Both features are similar to polyps.

In Fig. 7(k), a large polyp is detected without false-negative. In this image, the colon is not well cleaned so that the image color is very different from the other most images. Two marks are correct on the same polyp.

In Section 3.3, the feature normalization is done across all features. This is choosed because the observation that this type of normalization can slightly reduce the false-positive rate.

The program is written on Matlab 2007b platform. It takes about 13 seconds in processing one colorectal image of the size 378×254 at a PC with a 1.83GHz Intel Centrino Duo CPU and 2GB RAM.

6 Conclusion

In this study, we have proposed a new feature set in detecting colorectal polyps on colonoscopic images. Seventy-four colonoscopic images are carefully selected containing a wide range of polyps types, polyps sizes, and different colors. Half of them are used as training images and half of them are used as test images. The support vector machine is utilized as a classifier. The result shows that the sensitivity reaches 86.2% and the false-positive rate is 1.26 per-image.

The result is promising and our future work is to combine the morphologic information with the texture information to get a more accurate detection result.

Acknowledgment

The authors would like to acknowledge China Medical University for supporting this research work under project CMU-95-280.

References

1. Cancer Facts and Figures. American Cancer Society (2003)
2. Rex, D., Weddle, R., Pound, D., O'Connor, K., Hawes, R., Dittus, R., Lappas, J., Lumeng, L.: Flexible sigmoidoscopy plus air contrast barium enema versus colonoscopy for suspected lower gastrointestinal bleeding. Gastroenterology 98, 855–861 (1990)
3. Taylor, S.A., Halligan, S., Slater, A., Goh, V., Burling, D.N., Roddie, M.E., Honeyfield, L., McQuillan, J., Amin, H., Dehmeshki, J.: Polyp detection with ct colonography: Primary 3d endoluminal analysis versus primary 2d transverse analysis with computer-assisted reader software. Radiology 239(3), 759–767 (2006)
4. Chowdhury, T.A., Whelan, P.F., Ghita, O.: The use of 3d surface fitting for robust polyp detection and classification in ct colonography. Computerized Medical Imaging and Graphics 30, 427–436 (2006)
5. Karkanis, S.A., Iakovidis, D.K., Maroulis, D.E., Karras, D.A., Tzivras, M.: Computer-aided tumor detection in endoscopic video using color wavelet features. IEEE Trans. on Information Technology in Biomedicine 7(3), 141–152 (2003)

6. Nagata, S., Tanaka, S., Haruma, K., Yoshihara, M., Sumii, K., Kajiyama, G., Shimamoto, F.: Pit pattern diagnosis of early colorectal carcinoma by magnifying colonoscopy: Clinical and histological implications. Int. J. Oncol. 16, 927–934 (2000)
7. Haralick, R.M., Shanmugam, K., Dinstein, I.: Textural features for image classification. IEEE Trans. on Systems, Man, and Cybernetics SMC-3(6), 610–621 (1973)
8. Cohen, P., Ledinh, C.T., Lacasse, V.: Classification of natural textures by means of two-dimensional orthogonal masks. IEEE Trans. on Acoustics, Speech, and Signal Processing 37(1), 125–128 (1989)
9. Esgiar, A.N., Naguib, R.N.G., Sharif, B.S., Bennett, M.K., Murray, A.: Automated feature extraction and identification of colon carcinoma. Anal. Quant. Cytology Histology 20, 297–301 (1998)
10. Esgiar, A.N., Naguib, R.N.G., Sharif, B.S., Bennett, M.K., Murray, A.: Microscopic image analysis for quantitative measurement and feature identification of normal and cancerous colonic mucosa. IEEE Trans. on Information Technology in Biomedicine 2, 197–203 (1998)
11. Tjoa, M.P., Krishnan, S.M.: Feature extraction for the analysis of colon status from the endoscopic images. BioMedical Engineering OnLine 2 (2003),
    http://www.biomedical--engineering--online.com/content/2/1/9
12. Iakovidis, D.K., Maroulis, D.E., Karkanis, S.A., Brokos, A.: A comparative study of texture features for the discrimination of gastric polyps in endoscopic video. In: Proc. of IEEE Symposium on Computer-Based Medical Systems (2005)
13. Conners, R.W., Harlow, C.A.: A theoretical comparison of texture algorithms. IEEE Trans. on Pattern Analysis and Machine Intelligence 2(3), 204–222 (1980)
14. Ohanian, P.P., Dubes, R.C.: Performance evaluation for four classes of textural features. Pattern Recognition 25(8), 819–833 (1992)
15. Walker, R.F., Jackway, P., Longstaff, I.D.: Improving co-occurrence matrix feaure discrimination. In: Proceedings of the 3rd Conference on Digital Image Computing: Techniques and Applications, pp. 643–648 (1995)
16. Matlab 2007b, the mathworks (2007), www.mathworks.com
17. Schölkopf, B., Smola, A.J.: Learning with Kernels. MIT Press, Cambridge (2002)

OplAnalyzer: A Toolbox for MALDI-TOF Mass Spectrometry Data Analysis

Thang V. Pham and Connie R. Jimenez

OncoProteomics Laboratory, Cancer Center Amsterdam,
VU University Medical Center
De Boelelaan 1117, 1081 HV Amsterdam, The Netherlands
{t.pham,c.jimenez}@vumc.nl
http://www.oncoproteomics.nl/

Abstract. We present a software package for the analysis of MALDI-TOF mass spectrometry data. The software is designed to facilitate a complete exploratory workflow: pre-processing of raw spectral data, specification of study groups for comparison, statistical differential analysis, visualization of peptide peaks, and classification. The software supports various external tools for these tasks. We also pay special attention to the iterative nature of a typical analysis. Finally, we present two proteomics studies where the software has been used for data analysis.

Keywords: data analysis, differential analysis, bio-marker discovery, MALDI-TOF, mass spectrometry, OplAnalyzer, proteomics.

1 Introduction

Mass spectrometry is an attractive method in proteomics research because of its ability to identify and quantify a large number of proteins in complex biological samples [1]. However, the pre-processing and analysis of mass spectrometry data are fast becoming a bottle neck in the discovery process. This paper describes a software platform developed in our laboratory called OplAnalyzer, which supports proteomics mass spectrometry data pre-preprocessing and analysis. Specifically, we deal with MALDI-TOF mass spectrometry, a standard high throughput platform that can potentially be used for various diagnostic purposes.

There are a number of tasks involved in a typical analysis: pre-processing of raw spectral data, specification of study groups for comparison, statistical differential analysis, visualization of peptide peaks, and classification [2]. Instead of integrating all these components into a single tool for a complete analysis, we develop a flexible platform where various existing tools for different tasks are accommodated. Our design also supports the interactive nature of the analysis process.

Currently, the software supports the analysis of MALDI-TOF MS-1 data only. Tools for the analysis of MS/MS data with protein identification as well as data from another mass spectrometry platform namely LC-FTMS are under active development.

P. Perner and O. Salvetti (Eds.): MDA 2008, LNAI 5108, pp. 73–81, 2008.
© Springer-Verlag Berlin Heidelberg 2008

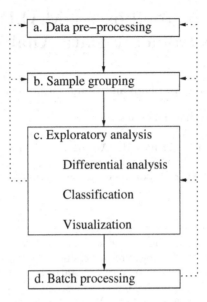

Fig. 1. An analysis workflow

The analysis workflow and the system are described in Section 2. In section 3 we present two proteomics studies where the software has been employed for data analysis.

2   The System

Fig. 1 shows a typical workflow in proteomics mass spectrometry data analysis. The four main steps are: data pre-processing, sample grouping, exploratory analysis, and batch processing.

2.1   Data Pre-processing

The data pre-processing step includes the preparation of metadata and the processing of raw mass spectrometry signals which consists of peak detection, alignment, normalization, and deisotoping. To facilitate the use of existing tools we define a common data format between this step and the subsequent steps, which is simply based on tab-separated texts.

For our instrument, a 4800 MALDI-TOF/TOF mass spectrometer (Applied Biosystems, Foster City, USA), we found that the MarkerView software (Applied Biosystems) works well for data produced in the reflectron mode.

For data produced in the linear mode we have implemented a new method. To detect peaks in an individual spectrum, we search for locations of maximal value within a local m/z window. The size of the window is 11 discrete sampling points. This method is similar to the peak detection method employed in [4].

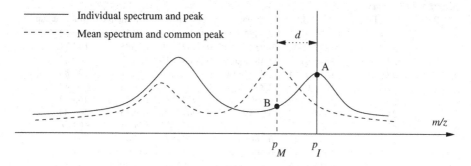

Fig. 2. Peak alignment. For each common peak p_M in the mean spectrum, the closest peak p_I in each individual spectrum is located. If the distance d between the two peaks is less than $\sqrt{5}$, the value at point A is registered for the common peak p_M in this particular spectrum. Otherwise, the value at B is registered.

To find peaks that are common in all spectra, we apply peak detection to the mean spectra, analogously to [5]. Subsequently, peaks in an individual spectrum are aligned to this set of common peaks as follows. For each common peak, its value in an individual spectrum is that of the closest detected peak in that spectrum if the distance between the common peak and the closest peak (in the m/z axis) is less than $\sqrt{5}$ Da. (A better choice is likely to be based on the actual mass accuracy of the measurement and on the m/z value.) If there is no such peak, the value is simply assigned to the value of the spectrum at the m/z location of the common peak. Figure 2 illustrates the procedure. By visual inspection, we found that the quality of our alignment method is comparable to that of the more computationally expensive clustering method in [4] (data not shown).

2.2   Sample Grouping

Typically, researchers are interested in several comparisons in each experiment, for examples, comparisons based on gender, age, and clinical outcomes. Also, in an interactive analysis the user might want to modify the sample groups for instance to include or exclude certain samples. To enable an efficient sample grouping, we define a text-based sample selection based on metadata. The strategy is easy to use and particularly suited for batch processing. For example, to specify two groups "Healthy" consisting of samples from healthy individuals and "Cancer" consisting of samples from cancer patients before treatment, the selection is as follows.

```
Healthy:Cancer-type=Healthy;Cancer:Cancer-type=NSCLC,Time=PreTx
```

Rank	Peak	Mz	p-value	fdr	f-change	max median
1	346	1777.966	2.1824e-006	0.00012222	3.0626	630.62
2	376	1877.9926	9.9568e-006	0.00027879	1.7685	147.71
3	275	1545.616	2.487e-005	0.00031645	2.2407	360.85
4	321	1690.9254	2.8254e-005	0.00031645	2.4294	272.85
5	373	1865.0022	2.8254e-005	0.00031645	3.9092	957.43
6	101	1039.6249	4.6693e-005	0.0004358	1.724	4690.2
7	213	1361.7417	5.9739e-005	0.00047791	1.5835	363.34
8	43	880.4127	7.6188e-005	0.00053331	1.9091	6022.6
9	102	1041.6349	0.00010908	0.00067871	1.786	3250
10	393	1955.9949	0.00013801	0.00077287	2.0098	124.35
11	93	1015.6267	0.00017407	0.00081232	1.699	11932
12	124	1087.5722	0.00017407	0.00081232	1.5078	809.97
13	112	1063.6231	0.00034269	0.0011994	1.6273	1691.3
14	126	1092.5826	0.00034269	0.0011994	1.5139	591.59
15	469	2318.2202	0.00034269	0.0011994	1.6218	113.02
16	507	2536.2991	0.00034269	0.0011994	1.737	491.79
17	422	2105.4972	0.00047573	0.0015671	2.0065	353.56
18	226	1396.5679	0.0005897	0.0018346	1.7299	236.76
19	423	2112.022	0.00072868	0.0021477	1.5405	80.503
20	80	987.5905	0.0013492	0.0037779	1.5889	1427

Fig. 3. A screenshot of the output of the statistical testing module

2.3 Exploratory Analysis

For data analysis we exploit existing tools in Matlab (The MathWorks, Inc). A typical first step is unsupervised analysis with principle component analysis (PCA) using all peptide intensities. Here all data points are projected onto a two or three-dimensional space for visualization. The projection does not use any information of group labels. The purpose is two-fold. First, one can observe if the data are clustered in a low dimensional space according to group labels. Second, one can detect possible outliers or unusual pattern in the data by visual inspection.

For differential analysis, we provide interfaces for the t-test, Mann-Whitney U test, Kruskal-Wallis test. The p-values can be adjusted for multiple testing. The peptides are further subjected to intensity filtering, requiring that the median intensity of at least one group must be greater than 80 units and the fold change of the median intensities of the two groups must be greater than 1.5. (The numbers can be tuned for each study). Fig. 3 depicts a screenshot of the result of a comparative study.

The candidate peaks are examined visually by spectra overlay. Again, we use the visualization capability of Matlab for this purpose.

Finally, we provide classification model selection with support vector machine [3]. A grid search method is used to find the optimal parameter values. For each value in the grid, the generalization error is estimated by either leave-one-out cross validation or repeatedly splitting the data into two partitions randomly, one for training and one for testing. The grid point with lowest estimated generalization error is selected as our model for classification.

2.4   Batch Processing

We consider batch processing an important step in data analysis, especially with regard to reproducibility of figures and other results. In addition, batch processing helps produce a large number of figures of peptide peaks in a convenient format for visual examination. Again, we make use of the scripting capability of Matlab for this purpose.

3   Examples

In the following, we describe two studies where the current software has been employed for data analysis.

3.1   Time-Course MALDI-TOF-MS Serum Peptide Profiling of Non-small Cell Lung Cancer Patients Treated with Bortezomib, Cisplatin and Gemcitabine

This study performs serum peptide profiling of non-small cell lung cancer (NSCLC) patients treated with gemcitabine, cisplatin and bortezomib combinations before, during, and at end of treatment to discover peptide patterns associated with treatment-related effects and clinical outcomes [7].

Fig. 4 shows a three-dimensional PCA plot of serum peptide spectra of 13 healthy individuals and the pre-treatment serum spectra of 27 NSCLC patients.

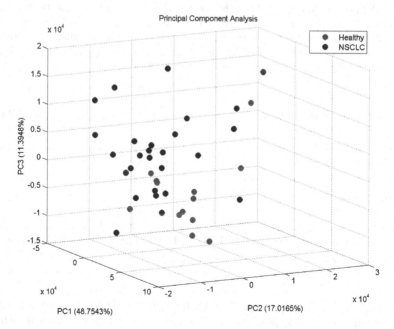

Fig. 4. Principle component analysis (PCA) of healthy versus NSCLC comparison

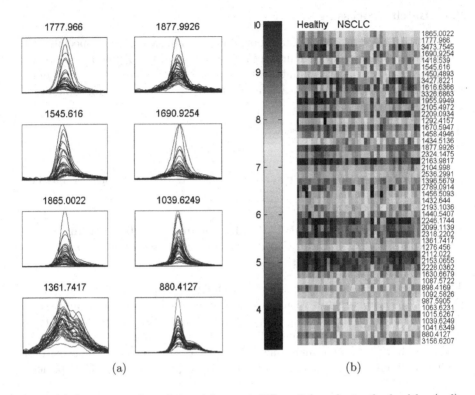

(a)   (b)

Fig. 5. (a) Spectra overlay of the eight most differential peaks in the healthy (red) versus NSCLC (blue) comparison according to p-values of the Mann-Whitney U test. All peaks have a p-value less than 0.0001. (b) Heatmap of the 47 differential peaks in the healthy versus NSCLC comparison shown in the natural log scale. The peaks are ordered by median fold change between the two groups.

Here, the MarkerView software was used for preprocessing, resulting in 682 peptide peaks per raw spectrum.

The Mann-Whitney U test is carried out on each of the 682 peptides, resulting in 47 differential peptides. Fig. 5(a) shows the spectra overlay of the eight most differential peaks in the healthy versus NSCLC comparison. Fig. 5(b) shows a heatmap of the 47 differential peaks.

We carried out classification analysis using support vector machine. A grid search for parameters was employed to find the best model according to leave-one-out cross validation (LOOCV). Using all 682 peptides, a LOOCV accuracy of 93% was achieved. When the 47 peptides selected by the Mann-Whitney U test were used, the LOOCV accuracy was 98% with 100% sensitivity and 96% specificity.

The software has also been used for a large number of other comparisons such as gender, age, short and long progression free survival, and clinical treatment responses.

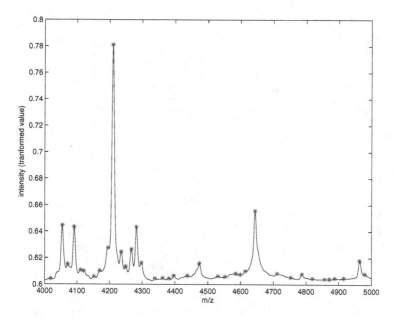

Fig. 6. Mean spectrum and detected peaks in the 4000-5000 Da range

3.2   Breast Cancer Study with Maldi-TOF Mass Spectrometry Data of Serum Samples

This study is part of the international competition on mass spectrometry proteomic diagnosis [8][9]. The dataset consists of 153 mass spectra of blood samples drawn from control individuals and patients with breast cancers. The aim is to construct a classification rule separating the two groups with a low generalization error.

For this dataset, the baseline correction had been performed by the competition organizer. We used the software to perform further pre-processing: peak detection and alignment. Fig. 6 shows an example of the result of the pre-procesing algorithm.

Again, a Mann-Whitney U test was performed to select features discriminating the two classes significantly. Furthermore, the Benjamini-Hochberg false discovery rate correction [6] was employed to correct for multiple testing. This results in on average 117 peaks with a false discovery rate less than 1%. Fig. 7 shows the distribution of the values of the 16 most discriminative peaks.

We employed grid search with exponential spacing to find the optimal values for support vector machine model selection. The generalization error is estimated by averaging over 200 runs of randomly splitting the given data into two partitions, where the size of the test set is roughly a tenth of size of the whole dataset. The feature selection was performed for each random splitting procedure, so that fair estimates of classification accuracy were obtained. The final accuracy on a separate validation set of 78 samples is 83%.

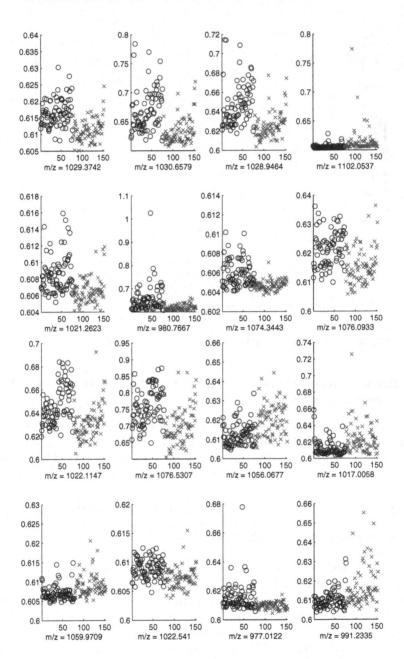

Fig. 7. Top 16 differential peaks

4 Summary

The paper has introduced a software toolbox for the pre-processing and statistical analysis of MALDI-TOF mass spectrometry data. Our current development focuses on the support for the analysis of MS/MS data with protein identification and data from another mass spectrometry platform namely LC-FTMS.

References

1. Jimenez, C.R., Piersma, S., Pham, T.V.: High-throughput and targeted in-depth mass spectrometry-based approaches for biofluid profiling and biomarker discovery. Biomarkers in Medicine 1(4), 541–565 (2007)
2. Villanueva, J., Martorella, A.J., Lawlor, K., Philip, J., Fleisher, M., Robbins, R.J., Tempst, P.: Serum peptidome patterns that distinguish metastatic thyroid carcinoma from cancer-free controls are unbiased by gender and age. Mol. Cell Proteomics 5, 1840–1852 (2006)
3. Vapnik, V.N.: The Nature of Statistical Learning Theory. Springer, Heidelberg (1999)
4. Tibshirani, R., Hastie, T., Narasimhan, B., Soltys, S., Shi, G., Koong, A., Le, Q.-T.: Sample classification from protein mass spectroscopy, by "peak probability contrasts". Bioinformatics 20(17), 3034–3044 (2004)
5. Karpievitch, Y.V., Hill, E.G., Smolka, A.J., Morris, J.S., Coombes, K.R., Baggerly, K.A., Almeida, J.S.: PrepMS: TOF MS data graphical preprocessing tool. Bioinformatics 23(2), 264–265 (2007)
6. Benjamini, Y., Hochberg, Y.: Controlling the false discovery rate: a practical and powerful approach to multiple testing. J. Roy. Statist. Soc. B 57, 289–300 (1995)
7. Voortman, J., Pham, T.V., Knol, J.C., Giaccone, G., Jimenez, C.R.: Time-course MALDI-TOF-MS serum peptide profiling of non-small cell lung cancer patients treated with bortezomib, cisplatin and gemcitabine. In: Proceedings of American Society of Clinical Oncology (ASCO) 2008 Annual Meeting, Chicago, USA (2008)
8. Mertens, B.: International competition on mass spectrometry proteomic diagnosis. Statistical Applications in Genetics and Molecular Biology 7(2), Article 1 (2008)
9. Pham, T.V., van de Wiel, M.A., Jimenez, C.R.: Support vector machine approach to separate control and breast cancer serum samples. Statistical Applications in Genetics and Molecular Biology 7(2), Article 11 (January 2008)

Classification of Mass Spectrometry Based Protein Markers by Kriging Error Matching

Tuan D. Pham[1], Honghui Wang[2], Xiaobo Zhou[3], Dominik Beck[1], Miriam Brandl[1], Gerard Hoehn[2], Joseph Azok[2], Marie-Luise Brennan[4], Stanley L. Hazen[4], and Stephen T.C. Wong[3]

[1] Bioinformatics Applications Research Center
James Cook University
Townsville, QLD 4811, Australia
[2] Clinical Center
National Institutes of Health
Bethesda, MD 20892, USA
[3] Department of Radiology
Weill Cornell Medical College and Methodist Hospital Research
Houston, Texas, USA
[4] Center for Cardiovascular Diagnostics and Prevention
Cleveland Clinic Foundation
Cleveland, OH 44195, USA

Abstract. Discovery of biomarkers using serum proteomic patterns is currently one of the most attractive interdisciplinary research areas in computational life science. This new proteomic approach has the clinical significance in being able to detect disease in its early stages and to develop new drugs for disease treatment and prevention. This paper introduces a novel pattern classification strategy for identifying protein biomarkers using mass spectrometry data of blood samples collected from patients in emergency department monitored for major adverse cardiac events within six months. We applied the theory of geostatistics and a kriging error matching scheme for identifying protein biomarkers that are able to provide an average classification rate superior to other current methods. The proposed strategy is very promising as a general computational bioinformatic model for proteomic-pattern based biomarker discovery.

Keywords: Bioinformatics; Pattern classification; Biomarker discovery; Proteomics; Mass spectrometry data; Geostatistics.

1 Introduction

A biomarker can be defined as an objectively measurable feature that indicates the current biochemical state of an individual [1]. These distinguishable states reveal the presence, absence or progress of a particular disease and therefore they can give insight into pharmacologic responses for therapeutic intervention [2]. Being used as diagnostic tools, biomarkers have the potential to personalize and

P. Perner and O. Salvetti (Eds.): MDA 2008, LNAI 5108, pp. 82–94, 2008.

improve medical treatment of patients on an individual basis. However, due to the molecular complexity of many diseases, paired with multiple phases through which they progress, discovery of the aforementioned biochemical features is still encountered with various types of challenges in both laboratory and analytical aspects. Therefore, the demand for highly specific and sensitive biomarkers has been continously increasing [2].

Despite the rapid progress in the field, there is no certain consensus about the best biomarker discovery strategy. Recent developments in biotechnology have provided the use of mass spectrometry (MS) as a very promising platform for the study of biomarker discovery. A particular advantage of MS over other protein profiling methods such as 2D gel electrophoresis is its high accuracy and resolution for even small molecular masses. Typical MS approach involves the analysis of samples collected from biofluids (for example, either serum or urine) or tissue of healthy and diseased patients. The molecules of these samples are first ionized (and hence charged), then separated after their molecular masses (m/z measured in Daltons), and finally the ion count is detected. This information is then translated into a typical MS spectrum. Regarding the large number of proteins in samples, an MS approach is generally coupled with a chromatographic step, which allows for the fractioning of the ions after molecular masses before they are processed via the MS. This process ensures a set of complexity reduced spectra. Recent MS studies have focused on the use of the Surface-enhanced laser desorption ionization (SELDI) method. For this method, samples are first added to a protein chip that incorporates some sort of affinity separation between noninformative and informative proteins [3]. Molecules with low affinity to the chip are washed away and immobilized proteins are ionized via the Matrix Assisted Laser Desorption Ionization (MALDI) method. Typically MALDI ions are then further studied by a time-of-flight (TOF) analyzer. The broad acceptance of SELDI MS for biomarker discovery comes mainly from its separation ability, which ensures comparable and complexity reduced spectra.

Once the mass spectra are generated, pattern classification methods can be utilized to identify the most discriminating peaks between diseased and control samples, which are considered as protein marker candidates. In regards to recent applications of proteomic technology, proteomic patterns have recently been utilized for early detection of cancer progressions [4,5,6], and computer methods for classification of normal and cancerous states using mass spectrometry data developed accordingly. Petricoin *et al.* [5] applied cluster analysis and genetic algorithms to detect early stage ovarian cancer using proteomic spectra. Ball *et al.* [7] applied integrated approach based on neural networks to study SELDI-MS data for classification of human tumors and identification of biomarkers. Lilien *et al.* [8] applied principal component analysis and a linear discriminant function to classify ovarian and prostate cancers. Sorace and Zhan [9] used mass spectrometry serum profiles to detect early ovarian cancer. Wu *et al.* [10] compared the performance of several methods for the classification of mass spectrometry data. Tibshirani *et al.* [11] proposed a probabilistic approach for sample classification from protein mass spectrometry data. Morris *et al.* [12] applied wavelet transforms and peak detection for feature extraction of MS data. Yu *et al.* [13] developed a method for dimensionality reduction for high-throughput MS data. Levner [14] used feature selection methods and then applied the nearest centroid technique to classify MS-based ovarian and prostate cancer datasets.

Given the promising integration of several machine-learning methods and mass spectrometry data in high-throughput proteomics [15], this new biotechnology still encounters several challenges in order to become a mature platform for clinical diagnostics and protein-based biomarker profiling. Some of major challenges include noise filtering of MS data, selection of computational methods for MS-based classification, feature extraction and feature reduction of MS datasets [16,17].

We have utilizied mass spectrometry data in the functional study of proteins for the prediction of the risk of major adverse cardiac events where advanced computational models play a critical role in the ability to analyze mass spectral data for risk prediction [18]-[22]. Our study was based the original work by Brennan *et al.* [23]. The authors of this paper studied 604 patients who presented in emergency room with chest pain. The blood samples were collected at the presentation of the emergency room and the protein level of MPO (myloperoxidase) and other known cardiovascular biomarkers were measured. The patient's outcome (any cardiovascular event) was monitored for 6 months. The study showed the MPO to be a new biomarker for the prediction of MACE (major adverse cardiac events) risk in 30 days after the presentation of chest pain in emergency room with accuracy about 60%. Recently, the FDA (U.S. Food and Drug Administration) approved the CardioMPO kit for measurement of MPO level (http://www.fda.gov/cdrh/reviews/K050029.pdf). As another study, we have recently applied the concept of distortion measures for classification of control and MACE samples [24]. However, this work did not address the identification of biomarkers for disease prediction. Another novel aspect of the current work is the introduction of the method of block kriging estimation error that serves as a basis for matching MS peaks, and hence identifying most informative biochemical features of MS data for early prediction of patient's risk of major adverse cardiac events.

The rest of this paper is organized as follows. Section 2 presents the kriging model and its error variance in the estimation of mass spectrometry data. Section 3 describes a classification scheme based on the concept of signal error matching. Experimental demonstrations and comparisons of the performances of the proposed and other methods are presented and discussed in Section 4. Section 5 concludes the findings and suggests potential issues for further investigation.

2   Error Estimation by Kriging

Geostatistics has been developed to describe the relationship of spatially distributed variables [25]. This type of statistical approach is embedded by the concept of regionalized variables. A regionized variable is defined as a random variable that is distributed in space. However, unlike random variables, regionalized variables have continuity from point to point but their changing behaviors are too complex to be expressed by any deterministic function. Therefore the spatial variability of the regionalized variables are considered to be both random and structured. This abstract notion is applicable to the modeling of mass spectrometry data where the samples can be thought as being both random and spatially related.

The computation of geostatistics involves the estimation of a regionalized variable in space. This estimation procedure is called kriging which is considered as a family of generalized linear regression methods [26,27]. Kriging estimates the value of a regionalized variable at a particular unsampled location by the weighted combination of the values of the neighboring locations. The kriging weights are derived using the information of the spatial covariance or the semi-variogram, where the latter function will be discussed in the next section. As a regionalized variable differs from a random variable, kriging differs from classical linear regression in that it neither assumes the variates are independent nor the observations are a random sample.

Let $s(n_k)$ be an MS peak at location n_k. Kriging can estimate the unknown MS value at a particular location n_0 as the linear combination of the known MS values at nearby locations as

$$\hat{s}(n_0) = \sum_{k=1}^{p} a_k s(n_k) \tag{1}$$

where a_k, $k = 1, \ldots, p$ are called the predictor coefficients or the kriging weights, and $s(n_k)$, $k = 1, \ldots, p$, are the known data values at locations n_k. The kriging weights are subject to

$$\sum_{k=1}^{p} a_k = 1 \tag{2}$$

The error variance of the kriging estimator can be defined as

$$\sum_{j=1}^{p} a_j C_{kj} + \beta = C_{k0}, \ \forall k = 1, \ldots, p. \tag{3}$$

where C_{kj} is the spatial covariance of $s(n_k)$ and $s(n_j)$, and β is a Lagrange multiplier.

The kriging system can be expressed in matrix notation as

$$\mathbf{C\,a} = \mathbf{b} \tag{4}$$

where $\mathbf{C}$ is the square and symmetrical matrix that represents the spatial covariances between the known MS peaks, and $\mathbf{b}$ is the vector that represents the spatial covariances between the unknown and known MS peaks:

$$\mathbf{C} = \begin{bmatrix} C_{11} & \cdots & C_{1p} & 1 \\ \cdot & \cdots & \cdot & \cdot \\ \cdot & \cdots & \cdot & \cdot \\ \cdot & \cdots & \cdot & \cdot \\ C_{p1} & \cdots & C_{pp} & 1 \\ 1 & \cdots & 1 & 0 \end{bmatrix}$$

$$\mathbf{a} = \begin{bmatrix} a_1 & \cdots & a_p & \beta \end{bmatrix}^T$$

$$\mathbf{b} = \begin{bmatrix} C_{10} & \cdots & C_{p0} & 1 \end{bmatrix}^T$$

Thus the vector of the spatial predictor coefficients can be obtained by solving

$$\mathbf{a} = \mathbf{C}^{-1} \mathbf{b} \tag{5}$$

The sample spatial covariance used for the kriging estimator can be calculated as [26]

$$C_{ij} = \frac{1}{N(h)} \sum_{(i,j)|h_{ij}=h} s(n_j) - \left(\frac{1}{n} \sum_{k=1}^{n} s(n_k)\right)^2 \tag{6}$$

in which the sample spatial covariance is a function of the lag distance h, $N(h)$ is the number of pairs that $s(n_i)$ and $s(n_j)$ are separated by h, and n is the total number of data.

Alternatively, the spatial covariances expressed in (4) can be replaced with the semi-variogram, denoted as $\gamma(h)$, for the determination of the spatial predictor coefficients. The experimental semi-variogram for lag distance h is defined as the average squared difference of values separated by h:

$$\gamma(h) = \frac{1}{2N(h)} \sum_{(i,j)|h_{ij}=h} [s(n_i) - s(n_j)]^2 \tag{7}$$

The properties of the semi-variogram can be explored by again letting h be the distance between two variables $s(n_i)$ and $s(n_j)$, and by an assumption that the random variables in the random function model has the same mean μ and variance σ^2. These two properties show the relationship between the semi-variogram and the covariance by the following derivation [26]:

$$\begin{aligned} \gamma(h) &= \frac{1}{2} E\{[s(n_i) - s(n_j)]^2\} \\ &= \frac{1}{2} E\{s(n_i)^2\} + \frac{1}{2} E\{s(n_j)^2\} - E\{s(n_i)s(n_j)\} \\ &= E\{s^2\} - E\{s(n_i)s(n_j)\} \\ &= [E\{s^2\} - \mu^2] - [E\{s(n_i)s(n_j)\} - \mu^2] \\ &= \sigma^2 - C_{ij} \end{aligned} \tag{8}$$

Using the semi-variogram function for solving the kriging system of equations, the error variance of estimation of the point or ordinary kriging can be determined as

$$\sigma_{OK}^2 = \mathbf{a}^T \mathbf{b} \tag{9}$$

What has been discussed is known as the point kriging procedure for point estimation. However, it is often required an estimate of the average value of a variable within a prescribed local area or volume. Block kriging has the capability of computing such an estimate in an effective way that can avoid the expense of computational effort in the averaging of point-by-point computing [26]. Being similar to point kriging, the block kriging system is expressed the same as that of the point kriging given in (4), but the $\mathbf{b}$ component, which is now denoted by $\mathbf{b}_{BK}$ to stand for block kriging:

$$\mathbf{C}\,\mathbf{a} = \mathbf{b}_{BK} \tag{10}$$

where $\mathbf{b}_{BK}$ is defined as the vector that represents the average spatial covariances between a particular sample location and all the points within a domain A:

$$\mathbf{b}_{BK} = \begin{bmatrix} C_{1A} \cdots C_{nA} \, 1 \end{bmatrix}^T$$

where the average spatial covariances between a sample location and all the points within A is defined as

$$C_{iA} = \frac{1}{N} \sum_{j|j\in A}^{N} C_{ij} \tag{11}$$

The block kriging estimation error variance is given by

$$\sigma_{BK}^2 = C_{AA} - \mathbf{a}^T \mathbf{b}_{BK} \tag{12}$$

where

$$C_{AA} = \frac{1}{NM} \sum_{i|i\in A}^{M} \sum_{j|j\in A}^{N} C_{ij} \tag{13}$$

In terms of the semi-variogram, where all the spatial covariances in $\mathbf{C}$ and $\mathbf{a}$ are replaced with the semi-variogram values, the block kriging error variance can be determined by

$$\sigma_{BK}^2 = \mathbf{a}^T \mathbf{b}_{BK} \tag{14}$$

3   Classification by Kriging Error Matching

Let $\mathbf{x}$, $\mathbf{y}$, and $\mathbf{z}$ be the vectors defined on a vector space V. A metric or distance d on V is defined as a real-valued function on the Cartesian product $V \times V$ if it has the properties of positive definiteness, symmetry, and triangle inequality. If a measure of dissimilarity satisfies only the property of positive definiteness, it is referred to as a distortion measure which is considered very common for the vectorized representations of signal spectra [28].

In general, to calculate a distortion measure between two vectors $\mathbf{x}$ and $\mathbf{y}$, denoted as $D(\mathbf{x}, \mathbf{y})$, is to calculate a cost of reproducing any input vector $\mathbf{x}$ as a reproduction of vector $\mathbf{y}$. Given such a distortion measure, the mismatch between two signals can be quantified by an average distortion between the input and the final reproduction. Intuitively, a match of the two patterns is good if the average distortion is small. A very useful distortion measure, derived from the theory of linear predictive coding (LPC), is the likelihood ratio distortion between the two templates presented in the form of two vectors of predictor coefficients $\mathbf{a}$, and $\mathbf{a}'$ which are used to model signal s. The likelihood-ratio distortion measure, denoted by D_{LR}, is defined as [28]

$$D_{LR}(\mathbf{a}, \mathbf{a}') = \frac{\mathbf{a}'^T \mathbf{R}_s \mathbf{a}'}{\mathbf{a}^T \mathbf{R}_s \mathbf{a}} - 1 \tag{15}$$

where $\mathbf{R}_s$ is the autocorrelation matrix of sequence s associated with its LPC coefficient vector $\mathbf{a}$, and $\mathbf{a}'$ is the LPC coefficient vector of signal s'. For a perfect match between the two templates, the errors are identical and (15) yields a zero distortion. For a mismatch, the residual resulting from the LPC analysis is large and the distortion defined in (15) therefore becomes large.

Based on the same principle derived for the likelihood ratio distortion, the block-kriging distortion measures, denoted as D_{BK}, can be defined as

$$D_{BK}(\mathbf{a}, \mathbf{a}') = \frac{\mathbf{a}'^T \mathbf{b}_{BK}}{\mathbf{a}^T \mathbf{b}_{BK}} - 1 \tag{16}$$

where $\mathbf{a}$ is defined in (4) which is the predictor vector of signal s, $\mathbf{b}_{BK}$ is the vector defined in (10) associated with s, and $\mathbf{a}'$ is the predictor vector of signal s'.

If the input (unknown) MS signal s_m is analyzed by the block kriging which results in a set of block kriging coefficients. The distortion measures between an MS peak of unknown class s_m and labeled samples can be determined. Using the best-match decision, the unknown signal s_m is assigned to class i^* if the distortion measure of its predictor vector $\mathbf{x}_m$ and the corresponding predictor vectors $\mathbf{c}^i$ is minimum, that is

$$s_m \rightarrow i^*, \quad i^* = \arg\min_i D_{min}(\mathbf{x}_m, \mathbf{c}^i) \tag{17}$$

where

$$D_{min}(\mathbf{x}_m, \mathbf{c}^i) = \min_j D(\mathbf{x}_m, \mathbf{c}^i_j) \tag{18}$$

where D is a distortion measure, $\mathbf{x}_m$ is the predictor vector of s_m, $\mathbf{c}^i_j$ is the predictor vector of the j sample that belongs to class i.

Another simple approach for classifying the unknown MS peak to either normal or diseased population is by using the majority vote; that is if the unknown

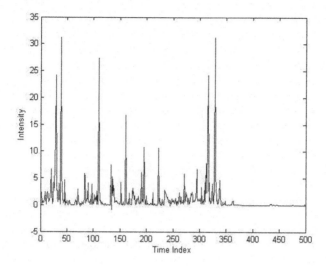

Fig. 1. SELDI-MS control sample

sample is labeled to one class more than the other, then sample is assigned to the class that has the major labeling:

$$\text{Assign } s_m \text{ to class } i^* \text{ if } V(i^*) > V(i), i^* \neq i, \forall i \tag{19}$$

where $V(i)$ is the number of votes for class i assigned by labeled samples based on the minimum distortion measure of the unknown sample with respect to the normal and diseased classes using (18).

4   Experimental Results

We used high-throughput, low-resolution SELDI MS (www.ciphergen.com) to acquire the protein profiles from patients and controls. Figures 1 and 2 show the typical SELDI mass spectra of the control and MACE samples respectively, where the m/z values are converted to time indices [18]. The protein profiles were acquired from 2 kDa to 200 kDa. The design of the experiment originally described in [18], which involves the datasets for the control and MACE group. Figures 3 is the box plotting of MS peak values at some typical time indices for both control and MACE samples. From the descriptive statistics of the box plot that graphically shows the summary of the smallest observation, lower quartile, median, upper quartile, and largest observation of each peak; it can be appreciated that the detemination of distinct peaks is a difficult task as the statistics of the two groups (control and MACE) is quite similar to one another.

For the control group, the dataset consists of sixty patients who presented in emergency room with chest pain and the patients' troponin T test was consistently negative. These patients lived in the next 5 years without any major cardiac events or death. The total 166 plasma samples, 24 reference samples and

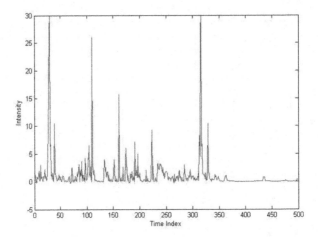

Fig. 2. SELDI-MS MACE sample

6 blanks were fractionated into 6 fractions using two 96-well plates containing anion exchange resin (Ciphergen, CA).

For the MACE group, the dataset was designed to comprise 60 patients who presented in emergency room with chest pain but the patients' troponin T test was negative. However, the patients in this group had either a heart attack, died or needed revascularization in the subsequent 6 months. The blood samples used in this study were same as those used in [23]. Most new MPO data measured with FDA approved CardioMPO kit for these two groups are available – MPO levels for 56 (out of 60) patients in control group and 55 (out of 60) patients in MACE group are available. Statistical analysis shows that MPO alone can distinguish MACE from control with an accuracy better than 60%.

For the SELDI mass spectra, the coverage of proteins in SELDI protein profiles was increased by that the blood samples were fractionated with HyperD Q (strong ion exchange) into 6 fractions. The protein profiles of fractions were acquired with two SELDI Chips: IMAC and CM10. There are a few different SELDI chips with different protein binding properties. Generally speaking, the more types of the SELDI chips are used, the more proteins are likely to be detected. However, due to the high concentration dynamic range of the proteins in human blood, the total number of proteins to be detected by the protocol we are using is very limited. We estimate that the number of the proteins we are able to detect is about one-thousand, while the total protein number in human blood is estimated to be tens of thousands. For example, MPO can be accurately measured with immunoassay (CardioMPO) but could not be detected with SELDI MS. The MS data for each sample in each fraction was acquired in duplicate, so 120 samples (60 controls and 60 MACEs) in each fraction in one type of SELDI chip have 240 spectra. There are two types of SELDI chips: IMAC and CM10.

We applied the block-kriging method for extracting the predictor coefficients and classifying control and MACE peaks. The variances of block-kriging estimation errors were determined by both left and right neighbors of the peaks where

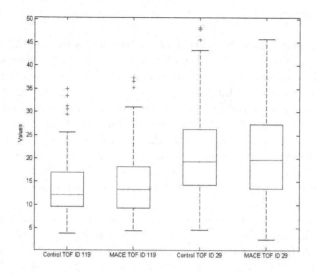

Fig. 3. Box plotting of some MS peaks of control and MACE at typical time indices

Table 1. Average classification rates by different methods

Method	Average accuracy (%)
MPO value	55.25
T-test	62.23
Standard genetic algorithm	69.05
Sequential forward floating search	71.92
Improved genetic algorithm	75.16
Block-kriging based best-match rule	67.95
Block-kriging based majority vote	93.15

the range of p was from 3 to 20 on each side. For the majority vote, we used an odd number of 3 peaks that have the minimum distortion with the unlabeled peak, to avoid a tie in assigning the class to the test samples. Performing the leave-one-out validation for different values of p, we obtained average classification rates of 67.95% and 93.15% using the best-match decision rule and majority vote, respectively.

In previous work [18], we used the MPO value, five selected biomarkers by T-test, five selected biomarkers by the sequential forward floating search, five selected biomarkers by standard genetic algorithm, and five selected biomarkers by an improved genetic algorithm to carry out the prediction. The average validation results of different methods are presented in Table 1 which shows the superior performance of the block-kriging based majority vote. The block-kriging based best-match scheme was found to be better than the use of the MPO and T-test; however, outperformed by the genetic algorithms and the searching technqiue.

5   Conclusion

We have presented a novel geostatistical strategy for identifying MS-based protein biomarkers. The proposed method is computationally feasible, mathematically tractable, and effective. It has the potential for applications to many other similar modern time-series biochemical or biological data, and hence would be very useful for new drug discovery. The classification rules we have addressed are simple methods in pattern recognition and classification. Our current intent is to demonstrate the advantage of the block-kriging method and its error matching scheme for distinguishing MS peaks from control and MACE samples. Improvements on the current results would be expected if other sophisticated classifiers can be explored for implementation using the same extracted features. These will be our future study.

It has been predicted that the advancement of proteomics pattern diagnostics might represent a revolution in the field of molecular medicine, because this technology has the potential of developing a new model for early disease detection [6,19,20,30]. Given that the research into clinical proteomic pattern diagnostics is still in its infancy because the results have not been validated in large trials, and effective computational methods have not been well-explored; recent research outcomes have illustrated the role of MS-based proteomics as an indispensable tool for molecular and cellular biology and for the emerging field of systems biology [29]. Nevertheless, research into MS-based disease detection has recently attracted the attention of researchers from various disciplines. In particular, it offers tremendous potentials for the development of personalized medicine [31] – a new concept that major diseases have a genetic component; therefore the understanding of cellular processes at the molecular level will enable scientists and physicians to predict the relative risk and potential therapy for such conditions on a person-to-person basis.

Furthermore, the challenge for the post-genomic era is to understand how genetic regulatory network interact as a system and how this system function creates an organism. This area of research is called *systems biology* [32]. Systems biology is still in the stage of infancy and therefore needs different tools from a variety of different perspectives. The novel applications of various computational methods for high-throughput data such as mass spectrometry and microarray gene expression data are important tools of systems biology and have proven invaluable for identifying and characterizing the components of biological systems on a comprehensive scale [32]. It is hoped that the method we developed herein would become one of such effective feature extraction and classification approaches of systems biology for both time-course and imaging data.

References

1. Veenstra, T.D.: Global and targeted quantitative proteomics for biomarker discovery. J. Chromatography B 847, 3–11 (2007)
2. Schrader, M., Selle, H.: The process chain for peptidomic biomarker discovery. Disease Markers 22, 27–37 (2006)

3. Diamandis, E.P.: Mass Spectrometry as a diagnostic and a cancer biomarker discovery tool: Opportunities and potential limitations. Mol. Cell Proteomics 3, 367–378 (2004)
4. Sauter, E., et al.: Proteomic analysis of nipple aspirate fluid to detect biologic markers of breast cancer. Br. J. Cancer 86, 1440–1443 (2002)
5. Petricoin, E.F., et al.: Use of proteomic patterns in serum to identify ovarian cancer. Lancet 359, 572–577 (2002)
6. Conrads, T.P., Zhou, M., Petricoin III, E.F., Liotta, L., Veenstra, T.D.: Cancer diagnosis using proteomic patterns. Expert Rev. Mol. Diagn. 3, 411–420 (2003)
7. Ball, G., et al.: An integrated approach utilizing artificial neural networks and SELDI mass spectrometry for the classification of human tumours and rapid identification of potential biomarkers. Bioinformatics 18, 395–404 (2002)
8. Lilien, R.H., Farid, H., Donald, B.R.: Probabilistic disease classification of expression-dependent proteomic data from mass spectrometry of human serum. J. Computational Biology 10, 925–946 (2003)
9. Sorace, J.M., Zhan, M.: A data review and re-assessment of ovarian cancer serum proteomic profiling. BMC Bioinformatics 4, 24 (2003)
10. Wu, B., et al.: Comparison of statistical methods for classification of ovarian cancer using mass spectrometry data. Bioinformatics 19, 1636–1643 (2003)
11. Tibshirani, R., et al.: Sample classification from protein mass spectrometry, by peak probability contrasts. Bioinformatics 20, 3034–3044 (2004)
12. Morris, J.S., Coombes, K.R., Koomen, J., Baggerly, K.A., Kobayashi, R.: Feature extraction and quantification for mass spectrometry in biomedical applications using the mean spectrum. Bioinformatics 21, 1764–1775 (2005)
13. Yu, J.S., Ongarello, S., Fiedler, R., Chen, X.W., Toffolo, G., Cobelli, C., Trajanoski, Z.: Ovarian cancer identification based on dimensionality reduction for high-throughput mass spectrometry data. Bioinformatics 21, 2200–2209 (2005)
14. Levner, I.: Feature selection and nearest centroid classification for protein mass spectrometry. BMC Bioinformatics 6, 68 (2005)
15. Shin, H., Markey, M.K.: A machine learning perspective on the development of clinical decision support systems utilizing mass spectra of blood samples. J. Biomedical Informatics 39, 227–248 (2006)
16. Anderle, M., Roy, S., Lin, H., Becker, C., Joho, K.: Quantifying reproducibility for differential proteomics: noise analysis for protein liquid chromatography-mass spectrometry of human serum. Bioinformatics 20, 3575–3582 (2004)
17. Salmi, J., Moulder, R., Filen, J.-J., Nevalainen, O.S., Nyman, T.A., Lahesmaa, R., Aittokallio, T.: Quality classification of tandem mass spectrometry data. Bioinformatics 22, 400–406 (2006)
18. Zhou, X., Wang, H., Wang, J., Hoehn, G., Azok, J., Brennan, M.L., Hazen, S.L., Li, K., Wong, S.T.C.: Biomarker discovery for risk stratification of cardiovascular events using an improved genetic algorithm. In: Proc. IEEE/NLM Int. Symposium on Life Science and Multimodality, pp. 42–44 (2006)
19. Petricoin, E.F., Liotta, L.A.: Mass spectrometry-based diagnostics: The upcoming revolution in disease detection. Clinical Chemistry 49, 533–534 (2003)
20. Wulfkuhle, J.D., Liotta, L.A., Petricoin, E.F.: Proteomic applications for the early detection of cancer. Nature 3, 267–275 (2003)
21. Goodacre, S., Locker, T., Arnold, J., Angelini, K., Morris, F.: Which diagnostic tests are most useful in a chest pain unit protocol? BMC Emergency Medicine 5, 6 (2005)
22. Wu, A.: Markers for Early Detection of Cardiac Diseases. Scandinavian Journal of Clinical and Laboratory Investigation suppl. 240, 112–121 (2005)

23. Brennan, M.-L., Penn, M.S., Van Lente, N.V., Shishehbor, M.H., Aviles, R.J., Goormastic, M., Pepoy, M.L., McErlean, E.S., Topol, E.J., Nissen, S.E., Hazen, S.L.: Prognostic value of myeloperoxidase in patients with chest pain. The New England Journal of Medicine 13, 1595–1604 (2003)

24. Pham, T.D., Wang, H., Zhou, X., Beck, D., Brandl, M., Hoehn, G., Azok, J., Brennan, M.-L., Hazen, S.L., Li, K., Wong, S.T.C.: Computational prediction models for early detection of risk of cardiovascular events using mass spectrometry data. IEEE Trans. Information Technology in Biomedicine (in print, 2007), doi:10.1109/TITB.2007.908756

25. Matheron, G.: The theory of regionalized variables and its applications. Paris School of Mines Publication, Paris (1971)

26. Isaaks, E.H., Srivastava, R.M.: An Introduction to Applied Geostatistics. Oxford University Press, New York (1989)

27. Davis, J.C.: Statistics and Data Analysis in Geology. John Wiley & Sons, New York (2002)

28. Rabiner, L., Juang, B.-H.: Fundamentals of Speech Recognition. Prentice Hall, New Jersey (1993)

29. Aebersold, R., Mann, M.: Mass spectrometry-based proteomics. Nature 422, 198–207 (2003)

30. Petricoin, E.F., Rajapaske, V., Herman, E.H., et al.: Toxicoproteomics: Serum proteomic pattern diagnostics for early detection of drug induced cardiac toxicities and cardioprotection. Toxicologic Pathology 32(suppl. 1), 1–9 (2004)

31. Ginsburg, G.S., McCarthy, J.J.: Personalized medicine: revolutionizing drug discovery and patient care. Trends Biotechnol. 19, 491–496 (2001)

32. Megason, S.G., Fraser, S.E.: Imaging in systems biology. Cell 130, 784–795 (2007)

A Mathematical Operator for Automatic and Real Time Analysis of Sequences of Vascular Images

Marcello Demi[1,2], Elisabetta Bianchini[1], Francesco Faita[1], and Vincenzo Gemignani[1]

[1] CNR Institute of Clinical Physiology, via G.Moruzzi 1, 56124 Pisa, Italy
[2] Esaote SpA, via di Caciolle 15, 50127 Firenze, Italy
{marcello.demi,elisabetta.bianchini,francesco.faita,
vincenzo.gemignani}@ifc.cnr.it

Abstract. Due to the absolute value involved, the first absolute central moment can be divided into two complementary filters: a positive deviation e_p and a negative deviation e_n. Both e_p and e_n can be used separately to highlight edges, lines, line endings, corners and junctions in images. Furthermore, the recovered edge information can be usefully combined to obtain additional information that would not be obtained by varying the parameters of the original filter. The mass center of the first absolute central moment can be also defined and an iterative localization procedure can be developed by exploiting its properties. Mathematical operators derived from the first absolute central moment were used on a video processing device based on a DSP board and they proved to be robust and suitable for real-time implementations.

Keywords: edge detection, contour tracking, ultrasound image sequences.

1 Introduction

The first absolute central moment belongs to the wide class of moments of n order, which includes variance, skewness and kurtosis [1-5]. However, the first absolute central moment has not been analyzed in depth in the past [6] and, in particular, its properties have never been exploited in image processing. It is common opinion that the first absolute central moment has not been investigated because of the mathematical difficulties introduced by the presence of the absolute value which makes theorem proving difficult [2][4]. Statistical measures such as median, mean, variance, skewness and kurtosis are often used in literature to describe the spatial features of a region of an image [3][5]. Mean and median are used as a measure of the luminance, variance is a measure of the width of the gray-level distribution, the skewness characterizes the degree of asymmetry of the distribution while the kurtosis is a measure of the peakedness or flatness of the distribution. Mean and median are also used to reduce noise which affects images [7]. The mean filter is used when the noise distribution well approximates a normal distribution. The median is used when the normal model would be a bad approximation of the noise distribution since the mean is not a robust estimator of the signal if pixels (outliers), with a gray level which is significantly different from the gray level of the surrounding pixels, are present [4].

P. Perner and O. Salvetti (Eds.): MDA 2008, LNAI 5108, pp. 95–107, 2008.
© Springer-Verlag Berlin Heidelberg 2008

Unlike the above statistical measures which are largely used in image processing systems, the first absolute central moment is only used in robust statistics. Here the median and the first absolute central moment are used to estimate the central value and the width, respectively, of the distribution since both these measures are robust against outliers [4] [6].

However, the first absolute central moment supplies features of the utmost importance in image processing. Due to the absolute value involved, the first absolute central moment can be separated into two components: a positive deviation e_p and a negative deviation e_n. Once e_p and e_n are computed they can be combined. Derivative filters of both the first order and the second order can be obtained as well as mechanisms which are able to compensate the noise effects. Edges can be located and image key points such as corners, lines, line-endings and intersections between different discontinuities can be highlighted [8][9]. The architectures of e_p and e_n and their responses to luminous stimuli recall the separation of the ganglion cells of a biological vision system into on-center off-surround and off-center on-surround cells [10].

When e_n and e_p are computed separately, the difference e_p-e_n is the first absolute central moment and the sum e_p+e_n represents the first central moment. That is, the central moment and the absolute central moment, can be obtained as combinations of e_p and e_n. While the first central moment is equivalent to a filter difference of Gaussians (DoG) [11][12], the first absolute central moment provides a result which is analogous to that provided by a filter gradient of Gaussian (GoG) [12][13]. Therefore, both a first derivative operator such as a GoG and a second derivative operator such as a DoG can be obtained with the same filtering stage. Moreover, unlike the GoG, the first absolute central moment provides ridges both at edges and at lines and highlights the image key points like corners, line-endings and junctions with local maxima. Notwithstanding, e_p and e_n represent something more than a simple trick to compute the two central and absolute central moments simultaneously: e_p and e_n are themselves two mathematical operators which reveal interesting properties. e_p and e_n also highlight edges and lines with a ridge and provide local maxima at the image key points. While e_n provides a ridge with the peak at the dark border of a gray-level discontinuity, e_p provides a ridge with the peak at the bright border. Moreover, the ridges provided by e_p and e_n at a discontinuity partially overlap, the profile of the overlapping area is that of a thin ridge and the peak of the ridge locates the discontinuity; the greater the discontinuity, the higher the peak. Consequently, a simple algebraic method can provide both the edge map and an estimate of the image luminance variation. The function *Min(positive deviation, |negative deviation|)* provides both a map M_{pn}, similar to the zero-crossing map of an equivalent DoG filter [14], and the strength of each zero-crossing point. In addition to the M_{pn} map a local thresholding procedure can be achieved by combining two ridge maps obtained when two different sets of low-pass filters are used.

Furthermore, other properties emerge from the analysis of the mass center of the first absolute central moment (vector **b**) [15]. If vector **b** is computed at a starting point p near an edge then vector **b** joins a point p' which is closer to the edge than p, independently of the distance between p and the edge. Therefore, an iterative localization procedure can be developed by exploiting this property. When given an approximate starting contour the final contour of the structure in interest can be located by computing iteratively vector **b** at the points of the starting contour. The procedure is simple and converges in just a few iterations. In this case also, due to the absolute value involved, vector **b** can be separated into two components: a positive component $\mathbf{b_p}$ and a negative component $\mathbf{b_n}$. Once $\mathbf{b_p}$ and $\mathbf{b_n}$ are computed they can be combined: the sum $\mathbf{b_p} + \mathbf{b_n}$ is the mass center of the first absolute central moment (that is, vector **b**) while the difference $\mathbf{b_p} - \mathbf{b_n}$ represents the normalized gradient of the gray-level image map.

Mathematical operators derived from the first absolute central moment were used on a stand-alone video processing device based on a DSP board where the main component is the Texas Instruments' TMS320C6415. The TMS320C6415 is a high performance digital signal processor particularly suited for computationally intensive video processing applications. The device was successfully used for the development of ultrasound image-processing applications. It acquires the analog video signal from any ultrasound system and shows the results on a Graphical User Interface (GUI). For example, the diameter of an artery can be estimated in real-time: for every image, that is, at a rate of 25 frames/sec, the DSP automatically locates the two borders of the vessel and subsequently computes the diameter. The method is based on a contour tracking procedure which exploits the properties of vector **b** and it is applied to B-mode images of a longitudinal section of the vessel. Long image sequences can be thus processed and complex exams like the analysis of the endothelial function can be performed in real time.

2   The First Absolute Central Moment

Let $f(n,m)$ be the gray-level map of an image and let Θ_1 and Θ_2 be two concentric circular neighborhoods of a point **p** with coordinates (n,m). Let r_1 and r_2, where $r_1 < r_2$, be the radii of Θ_1 and Θ_2, respectively. The first absolute central moment can be computed as follows:

$$e(n,m) = \sum\sum_{\Theta_2} |\mu(n,m) - f(n-k,m-l)| g_2(k,l) \tag{1}$$

where the mean value $\mu(n,m)$ is computed as

$$\mu(n,m) = \sum\sum_{\Theta_1} f(n-k,m-l) g_1(k,l) \tag{2}$$

and $g_1(k,l)$ and $g_2(k,l)$ are two Gaussian weight functions with apertures σ_1 and σ_2. Even if simpler weight functions like box functions could be used, we prefer Gaussian functions since Gaussian has many qualities which make this function a unique operator in early image processing [16-19]. In order to normalize the operator, the discrete Gaussian functions $g_1(k,l)$ and $g_2(k,l)$ are normalized over the circular neighborhoods Θi with radius $r_i = 3\sigma_i$. Eq.(1) measures the variability of the gray levels of the pixels which belong to the circular neighborhood Θ_2 of the image with respect to the local mean computed on the smaller circular neighborhood Θ_1. The first absolute central moment $e(n,m)$ can be divided into two complementary filters: a positive deviation $e_p(n,m)$ and a negative deviation $e_n(n,m)$.

$$e_p(n,m) = \sum\sum_{\Theta_{2p}} (\mu(n,m) - f(n-k,m-l))g_2(k,l)$$

$$e_n(n,m) = \sum\sum_{\Theta_{2n}} (\mu(n,m) - f(n-k,m-l))g_2(k,l)$$
(3)

where the domains Θ_{2p} and Θ_{2n} are defined as

$$\Theta_{2p} = \{(k,l) \in \Theta_2 : \mu(n,m) > f(n-k,m-l)\}$$
(4)

$$\Theta_{2n} = \{(k,l) \in \Theta_2 : \mu(n,m) < f(n-k,m-l)\}$$

so that the first absolute central moment can be obtained as $e(n,m)=e_p(n,m)-e_n(n,m)$. Given the local mean $\mu(n,m)$ computed on the smaller circular neighborhood Θ_1 eqs.(3) measure the variability of the gray levels of the pixels, which belong to the circular neighborhood Θ_2 and which are greater or less than $\mu(n,m)$, respectively, with respect to $\mu(n,m)$ itself.

3   The Positive and Negative Deviations

Fig.1 panel a) shows the response of both $e_p(n,m)$ and $e_n(n,m)$ given a test image with an ideal straight step discontinuity. At gray-level discontinuities $e_p(n,m)$ and $e_n(n,m)$ provide two ridges which overlap partially. The overlapping area is a thin ridge with a base equal to $2r_1$ where r_1 is the radius of Θ_1. While $e_n(n,m)$ provides a ridge with the peak at the dark border of the gray-level discontinuity, $e_p(n,m)$ provides a ridge with the peak at the bright border.

In general, the negative deviation highlights dark structures on bright backgrounds and vice versa, the positive deviation highlights bright structures on dark backgrounds. According to this property, $e_p(n,m)$ and $e_n(n,m)$ can be used to highlight the outer and inner border of structures in interest separately as well as both the borders and the center lines of bars. Fig.1 then shows how the outputs of $e_p(n,m)$ and $e_n(n,m)$ can be combined to obtain different results at discontinuities.

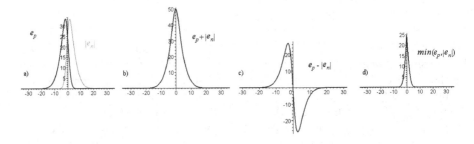

Fig. 1. The figure shows how the outputs of $e_p(n,m)$ and $e_n(n,m)$ can be combined to obtain different results at gray-level discontinuities

3.1 A Filter Analogous to a GoG Filter

When we subtract the output of $e_n(n,m)$ from $e_p(n,m)$ the output of the first absolute central moment $e(n,m)$ is obtained. While $e(n,m)$ is equal to zero over homogeneous regions, at discontinuities $e(n,m)$ provides a ridge and the ridge peaks locate the points of the discontinuity. The first absolute central moment is a dispersion index which provides a ridge map similar to the ridge map provided by the magnitude of a standard GoG operator.

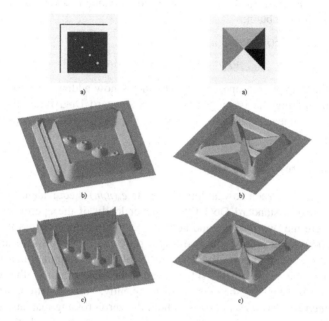

Fig. 2. Panels a) shows two test images. The relative ridge maps provided both by the first absolute central moment and by an equivalent GoG filter are shown in panels c) and b), respectively. Unlike the magnitude of the GoG, the first absolute central moment provides ridges at edges and lines and gives rise to local extrema of the ridges at line endings, corners, spots and junctions.

Fig.1 panel b) shows how the linear combination $e_p(n,m) - e_n(n,m)$ provides a ridge at discontinuities. However, unlike the magnitude of the GoG, the first absolute central moment provides ridges at edges and lines and gives rise to local extrema of the ridges at key points such as line endings, corners, spots and junctions [8]. Fig.2 shows two test images in panels a) and the relative ridge maps provided both by the first absolute central moment (panels c)) and by an equivalent GoG filter (panels b)). However, it is worth noting that the first absolute central moment cannot highlight particular junctions with a local maximum. Local maxima at all the junctions can be ensured only if the two positive and negative components (e_p and e_n) of the first absolute central moment are kept separated [9].

3.2 The Same Output of a Standard DoG Filter

The first central moment can be obtained by adding the negative deviation to the positive deviation. It is worth noting, however, that in our case the radius of Θ_1 is smaller than the radius of Θ_2 and the obtained operator $c(n,m)$ is a generalized first central moment rather than a real first central moment.

$$c(n,m) = \sum\sum_{\Theta_2} (\mu(n,m) - f(n-k,m-l))g_2(k,l) \tag{5}$$

Since $c(n,m)$ is obtained by eliminating the absolute value brackets in eq.(1), then by developing eq.(5) we obtain:

$$c(n,m) = e_p(n,m) + e_n(n,m) =$$
$$= f(n,m) \otimes (g_1(n,m) - g_2(n,m)) \tag{6}$$

where $\otimes$ is the convolution operator. Eq.(6) shows how adding $e_p(n,m)$ to $e_n(n,m)$ is equivalent to filtering the image $f(n,m)$ with a standard DoG filter. Fig.1 panel c) shows how the linear combination $e_p(n,m)+e_n(n,m)$ provides a zero-crossing at discontinuities.

3.3 A Simple Zero-Crossing Map

In section 3.2 we have seen how adding $e_p(n,m)$ to $e_n(n,m)$ is equivalent to filtering the image $f(n,m)$ with a standard DoG filter independently of the apertures σ_1 and σ_2. However, where the two positive and negative DoG components cross the zero (that is, at the zero-crossing points), the two ridges provided by the positive and negative central deviations overlap partially. The profile of the overlapping area is that of a thin ridge and the peak of the ridge locates the discontinuity; the greater the discontinuity, the higher the peak. Consequently, the function *Min(positive deviation,* |*negative deviation*|*)* provides both the zero-crossing map and an estimate of the image luminance variation (the zero-crossing strength). Fig.1 panel d) shows how the function M_{pn} provides a thin ridge at discontinuities. A similar procedure has been used in the past in [20] to reduce the noise effects on edge detection and localization.

3.4 A Local Thresholding Procedure

A local thresholding procedure can be achieved by combining two ridge maps obtained when two different sets of low-pass filters are used. The difference in noise compensation given by the two following filtering processes has been analyzed in [9]:

$$e_1(n,m) = \sum\sum_{\Theta_2} |\mu(n,m) - f(n-k,m-l)|g_2(k,l)$$
$$\mu(n,m) = \sum\sum_{\Theta_1} f(n-k,m-l)g_1(k,l) \tag{7}$$

$$e_2(n,m) = g_3(n,m) \otimes \sum\sum_{\Theta_2} |f(n,m) - f(n-k,m-l)|g_2(k,l) \tag{8}$$

Provided that $\sigma_1=\sigma_3$, eq.(7) provides a lower noise level than eq.(8). The opposite result, however, is obtained at the discontinuities. While in eq.(8) the height of the ridges at the discontinuities obviously decreases by filtering the central deviation $e(n,m)$ with the Gaussian $g_3(n,m)$, in eq.(7) the height of the ridges at the discontinuities does not decrease when the Gaussian $g_1(n,m)$ is used.

Fig. 3. Four test images and the result of a thresholding process is shown. Since eq.(7) produces both higher ridges and a lower noise level than eq.(8), then the map obtained by the filtering process (8) can be used as a threshold map of the ridge map obtained by the filtering process (7).

Therefore, as eq.(7) produces both higher ridges and a lower noise level than eq.(8), the map obtained by the filtering process (8) can be used as a threshold map of the ridge map obtained by the filtering process (7). Fig.3 shows four test images and the result of the thresholding process. In particular, the first panel shows the robustness of the thresholding process to additive Gaussian noise. The ridge map of the synthetic image was obtained with $\sigma_1=1$; $\sigma_2=4$; $\sigma_3=0$. Smaller values of σ_2 were used to compute the ridge maps of the outdoor images and the apertures $\sigma_1=1.3$; $\sigma_2=1.3$; $\sigma_3=0$ were used. The threshold map of the synthetic image was obtained with $\sigma_1=0$; $\sigma_2=4$; $\sigma_3=4$. The threshold maps of the outdoor images were obtained with the apertures $\sigma_1=0$; $\sigma_2=1.3$; $\sigma_3=1.3$.

4   The Mass Center of the Gray-Level Variability

The first absolute central moment is a statistical filter which measures the variability of the gray levels of the image with respect to the local mean. The function $h(\mathbf{p},k,l)$

$$h(\mathbf{p},k,l)=\left|\mu(\mathbf{p})-f(n-k,m-l)\right|g_2(k,l) \qquad (9)$$

which can be found in the integral (1) describes the spatial distribution of the variability of the gray levels with respect to the local mean computed at point $\mathbf{p}=(n,m)$.

The function $h(\mathbf{p},k,l)$ can be seen as a mass function which associates a mass value to every pixel surrounding $\mathbf{p}$ and the first absolute central moment $e(\mathbf{p})$ can be seen as the total mass of the variability of the gray levels at point $\mathbf{p}$. Therefore, the center of mass of the gray-level variability at point $\mathbf{p}$ is computed with the vector $\mathbf{b}(\mathbf{p})$:

$$\mathbf{b}(\mathbf{p}) = \frac{1}{e(\mathbf{p})}\sum\sum_{\Theta_2} h(\mathbf{p},k,l)\Gamma \quad \text{if } e(\mathbf{p}) \neq 0, \quad \mathbf{b}(\mathbf{p}) = 0 \quad \text{if } e(\mathbf{p}) = 0 \tag{10}$$

where Γ is a vector which has $-k,-l$ components.

4.1 An Edge Localization Procedure

Let us consider a gray-level discontinuity and a point $\mathbf{p}_0$ close to the discontinuity. Vector $\mathbf{b}$ always indicates the discontinuity. Moreover, when particular configurations of eq.(10) are chosen, vector $\mathbf{b}$ locates a point $\mathbf{p}_1$ which is closer to the discontinuity than $\mathbf{p}_0$ independently of the distance between $\mathbf{p}_0$ and the discontinuity. Hence, given a starting point the closest point of a discontinuity can be located by iteratively computing vector $\mathbf{b}$. Let $|\varepsilon|$ be the distance of $\mathbf{p}$ from a straight discontinuity, vector $\mathbf{b}(\mathbf{p})$ was computed and the following relationship was obtained

$$b(\varepsilon) = -\frac{\sigma_2\sqrt{2}e^{-\frac{\varepsilon^2}{2\sigma_2^2}}}{\sqrt{\pi}} \cdot \frac{erf\left(\frac{\varepsilon\sqrt{2}}{2\sigma_1}\right)}{1 - erf\left(\frac{\varepsilon\sqrt{2}}{2\sigma_1}\right)erf\left(\frac{\varepsilon\sqrt{2}}{2\sigma_2}\right)} \tag{11}$$

The symbolic computation system Maple V [21] was used in [22] to compute and analyze eq.(11). Vector $\mathbf{b}$ always indicates the discontinuity and its magnitude is symmetric with respect to the discontinuity. Fig.4 shows how the magnitude of the vector $\mathbf{b}(\varepsilon)$ varies for positive values of ε when $\sigma_2 = 4\pi$ pixels. The mass center computed at $\mathbf{p}$ approaches the discontinuity independently of the distance between $\mathbf{p}$ and the discontinuity if $|\mathbf{b}(\varepsilon)| < 2|\varepsilon|$ for every value of ε. From eq.(11) it is easy to

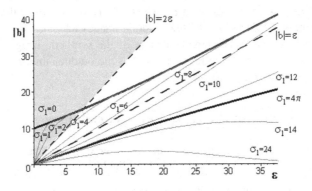

Fig. 4. The figure shows how the magnitude of the vector $\mathbf{b}(\varepsilon)$ varies for positive values of ε when $\sigma_2 = 4\pi$ pixels. The condition $\sigma_1 > \sigma_2/\pi$ is a necessary condition to satisfy the inequality $|b(\varepsilon)| < 2\varepsilon$ for every positive value of ε.

show that the magnitude of vector **b** decreases when σ_1 increases, independently of the values of ε and σ_2, and that the condition $\sigma_1 > \sigma_2/\pi$ is a necessary condition to satisfy the inequality $|b(\varepsilon)| < 2\varepsilon$ for every positive value of ε. Therefore, when the first absolute central moment is used, the mass center of the gray-level variability always approaches the discontinuity if the relationship $\sigma_1 > \sigma_2/\pi$ is satisfied.

4.2 An Operator Analogous to a Normalized Gradient

Here again the two positive and negative components of vector **b** can be introduced

$$
\mathbf{b_p(p)} = \frac{1}{e(\mathbf{p})} \sum\sum_{\Theta_{2p}} h(\mathbf{p},k,l)\Gamma \quad \text{if } e(\mathbf{p}) \neq 0, \quad \mathbf{b_p(p)} = 0 \quad \text{if } e(\mathbf{p}) = 0
$$

$$
\mathbf{b_n(p)} = \frac{1}{e(\mathbf{p})} \sum\sum_{\Theta_{2n}} h(\mathbf{p},k,l)\Gamma \quad \text{if } e(\mathbf{p}) \neq 0, \quad \mathbf{b_n(p)} = 0 \quad \text{if } e(\mathbf{p}) = 0
$$

(12)

where the domains Θ_{2p} and Θ_{2n} are defined in eqs.(4). Vector **b** is obviously obtained by adding the negative component $\mathbf{b_n}$ to the positive component $\mathbf{b_p}$ ($\mathbf{b(p)}=\mathbf{b_p(p)}+\mathbf{b_n(p)}$). However, a different operator $\mathbf{b_g}$ is obtained by subtracting $\mathbf{b_n}$ from $\mathbf{b_p}$: that is, by eliminating the absolute value brackets in eq.(9).

$$
\mathbf{b_g(p)} = -\frac{1}{e(\mathbf{p})} \sum\sum_{\Theta_2} f(n-k,m-l)\Gamma g_2(k,l) \quad \text{if } e(\mathbf{p}) \neq 0, \quad \mathbf{b_g(p)} = 0 \quad \text{if } e(\mathbf{p}) = 0 \quad (13)
$$

The summation in eq.(13) represents the convolution of the function $f(n,m)$ with the gradient of the Gaussian function $g_2(n,m)$ since, except for a constant factor, the term $\Gamma g_2(k,l)$ is exactly the gradient of the function $g_2(k,l)$. On the other hand, as we have seen in section 3.1, the normalizing factor $e(\mathbf{p})$ is very similar to the magnitude of the gradient of Gaussian. Therefore, given a gray-level image map $f(n,m)$, vector $\mathbf{b_g}$ provides the same information of a normalized gradient of Gaussian. A normalized gradient is needed, for example, to derive the velocity vector from the optical-flow equation.

5 The Real-Time Image Processing System

The video processing system we used to implement real-time measurements is a stand-alone video processing board. The main component is the Texas Instruments' TMS32C6415, a high performance DSP which is particularly suited for video processing applications. Its CPU, which has a Very-Long-Instruction-Word (VLIW) architecture, can carry out eight 32-bit instructions/cycle at 600MHz clock rate, that is 4.8 billion instructions/second. Moreover, special instructions can perform two arithmetic operations in parallel with 16-bit operands or four arithmetic operations with 8-bit operands, reaching a total of twenty-eight operations/cycle. The capability of performing multiple operations is of particular relevance in video processing applications since the gray-levels of the image are usually represented by 8-bit or 16-bit data. The board is equipped with an analog video decoder which can acquire the most common video standards, a multi-channel analog I/O module, a USB interface, an RS232 serial interface and a considerable amount of memory: 512 Kbytes flash memory; 4 Mbytes synchronous SRAM and 512 Mbytes DRAM. Unlike most common DSP video processing boards, which have a standard video output, our system is provided with a powerful VGA output generated by the Asiliant's B69000 graphics controller. This device integrates 2 Mbytes SDRAM, a 2D graphics accelerator and other hardware resources commonly available in PCs, such as a pop-up window and two cursors on the same chip. Moreover, high-quality video playback, which supports both a RGB and Y/C video format and implements a double buffering to eliminate video tearing, is obtained by using the B69000. Since the application requires advanced user interaction, a graphical user interface (GUI), which is operated by a mouse and a keyboard, has been developed. Although less complex, the GUI is similar to those we commonly use in standard workstations and can contain a number of objects such as buttons, numerical displays, texts, graphs etc. A more detailed description of the hardware and software architecture of the board can be found in [23].

6   Ultrasound Image Processing Examples

The system was used to develop novel ultrasound image processing applications. In these cases, the DSP board acquires the analog video signal from an ultrasound system and shows the results on a graphical user interface. A mouse and a keyboard are available to operate the device.

6.1   The Flow-Mediated Vasodilation

The endothelium is the tissue that lines the lumen of all blood vessels but overall it is an organ that synthesizes and releases vasoactive substances which regulate vascular functions. A dysfunction of this organ is an early step in the development of atherosclerosis and is a useful indicator for the prediction of cardiac events. For these reasons, the characterization of the endothelial function is one of the most attractive research topics in modern vascular medicine. The evaluation of the flow-mediated vasodilation (FMD) of the brachial artery is a widely used measurement technique. The examination consists in the application of a mechanical stimulus which results in the endothelium releasing nitric oxide, a vasodilator. Ultrasound equipment is used to measure the increasing of the artery diameter during the exam and, consequently, the response of the endothelium to the stimulus [24].

The system we developed can measure the diameter of the artery in real-time: for every image, that is, at a rate of 25 frames/sec, the DSP automatically locates the two borders of the vessel and subsequently computes the diameter. The method is based on a contour tracking technique applied to B-mode images of a longitudinal section of the vessel. Detecting the borders of the vessel may prove to be difficult because of the limited quality of the images and because of the presence of speckle noise. We approached the problem by exploiting the properties of vector **b** which proved to be robust and suitable for real-time implementations.

The procedure must be initialized by manually tracing two approximated starting borders, an operation that also defines the region of interest (ROI) where the diameter will be computed. This operation is carried out in a GUI window where the ultrasound images are displayed in real-time. The elaboration starts immediately afterwards. The value of the diameter is shown by a numerical display and is plotted on two graphs: the first graph shows the instantaneous measure over a time scale of 5 seconds; the second graph shows the mean value of the diameter computed over 2 seconds and displayed over a time scale of 9 minutes. The hardware device can be easily connected to any ultrasound equipment provided with an analog video output. At present 20 European research centers are successfully using the method with different ultrasound systems.

6.2 Carotid Intima-Media Thickness

Increased Carotid Intima-Media Thickness (CIMT) is a non-invasive marker of early arterial walls alteration which is associated with an increased risk for cardiovascular diseases. It can be easily assessed by a B-mode ultrasound technique and represents a safe and inexpensive measure which is well suited for use in large-scale population studies [25]. In particular, CIMT measurements are currently used as a cardiovascular (surrogate) end-point in randomized controlled trials with the advantage of reducing the sample size and the duration of follow-up [26].

CIMT is defined as the distance between the leading edge of the lumen-intima interface and the leading edge of the media-adventitia interface. To locate such edges, a manual approach is usually adopted. However, such a method is time consuming and rather unreliable, since results may depend on the experience, training and subjective judgment of the operator, thus involving a great inter- and intra-operator variability.

An automatic technique for the CIMT measurement, based on the first absolute central moment and a pattern recognition approach, have been developed. The mathematical operator exhibits an improved signal noise ratio in presence of speckle noise with respect to traditional edge detectors like Laplacian of Gaussian and Gradient of Gaussian, thus obtaining a greater precision in the CIMT assessment. Moreover, real-time visualization of the measure brings other benefits. Firstly, the time required for examination is very low. Secondly, physicians can take advantage of the visual feedback to adjust the quality of ultrasound images so to increase the global robustness of the measure.

7 Conclusion

The most interesting feature of the first absolute central moment is that the generalization of this simple dispersion index gives rise to a class of filters, the outputs of which can be in turn usefully combined. The first central moment, which is obtained by adding the negative deviation to the positive deviation, provides a map which is equal to the one provided by a standard DoG filter. The simple algebraic function *Min(positive deviation, |negative deviation|)* can also provide both the zero-crossing map and an estimate of the image luminance variation. Moreover, a simple thresholding procedure can be obtained by combining two maps provided by two different filtering processes. The first absolute central moment, which is obtained by subtracting the negative deviation from the positive deviation, provides a ridge map which is similar to the one provided by the GoG filter. However, unlike the GoG, the absolute moment provides ridges at edges and lines and gives rise to local extrema of the ridges at line endings, corners and junctions. In the presence of noise, due to the weak ridge produced at the two-dimensional discontinuities, the GoG may ensure a ridge at straight discontinuities yet may not ensure a ridge at corners, at junctions or in the proximity of borders with high curvature, thus making the localization of these points unreliable [27][28]. However, it is worth noting that even though the first absolute central moment at junctions always ensures a ridge comparable to the ones generated at the straight discontinuities belonging to the junction, it does not provide local maxima at all the junctions. Local maxima at junctions are ensured only if the two positive and negative deviations are kept separated. In addition, given a starting point, the closest point of a discontinuity can be located with just a couple of jumps by iteratively computing the mass center of the first absolute central moment (vector **b**). Finally, an operator very similar to a normalized gradient of the gray-level image map is obtained when subtracting the negative component of vector **b** from its positive component.

Both the first absolute central moment and its mass center were used on a video processing device based on a DSP board and they proved to be robust and suitable for real-time implementations.

To conclude, we wish to point out that all the possible combinations of the edge information, recovered by the class of filters generated by the generalization of the first absolute central moment, have not been analyzed. Since three different low-pass filters can be introduced into two nonlinear filters (positive and negative deviations) a large class of filters is obtained and possible combinations of the recovered edge information still remain to be investigated. The edge detection properties of the two negative and positive deviations when used separately also remain to be investigated in depth. These two operators should be analyzed separately and their analysis should be compared with the knowledge achieved on the low level stages of the biological vision systems since their architectures and their responses to luminous stimuli recall the separation of the ganglion cells into on-center off-surround and off-center on-surround cells.

References

1. Papoulis, A.: Probability, Random Variables, and Stochastic Process. Mc Graw-Hill, New York (1965)
2. Bevington, P.R.: Data reduction and error analysis for the Physical Sciences. McGraw-Hill, New York (1969)
3. Jain, A.K.: Fundamentals of digital image processing. Prentice-Hall, Englewood Cliffs (1989)
4. Press, W.H., Flannery, B.P., Teukolsky, S.A., Vetterling, W.T.: Numerical Recipes. Cambridge University Press, Cambridge (1991)
5. Pratt, W.K.: Digital image processing. Wiley, New York (1991)
6. Huber, P.J.: Robust Statistics. Wiley, New York (1981)
7. Chellappa, R.: Digital image processing. Computer Society press, Los Alamitos (1992)
8. Demi, M.: Contour Tracking by Enhancing Corners and Junctions. Computer Vision and Image Understanding 63, 118–134 (1996)
9. Demi, M., Paterni, M., Benassi, A.: The First Absolute Central Moment in Low-Level Image Processing, Comput. Vision Image Understanding 80, 57–87 (2000)

10. Demi, M.: An Artificial Vision Model Based on Statistical Filters. In: Proc. of the Brain-Machine Workshop, pp. 37–44 (2000)
11. Marr, D.: Vision. W.H.Freeman, San Francisco (1982)
12. Torre, V., Poggio, T.: On edge detection. IEEE Trans. Pattern Anal. Machine Intell. PAMI-8, 147–163 (1986)
13. Canny, J.: A computational approach to edge detection. IEEE Trans. Pattern Anal. Machine Intell. PAMI-8, 679–698 (1986)
14. Marr, D.C., Hildreth, E.C.: Theory of edge detection. In: Proc. Roy. Soc. London B, vol. 207, pp. 187–217 (1980)
15. Demi, M.: The First Absolute Central Moment as an Edge Detector. Journal of Nonlinear Analysis 47(9), 5815–5826 (2001)
16. Witkin, A.P.: Scale-space filtering. In: Proc. Int. Joint Conf. Artificial Intelligence, pp. 1019–1022 (1983)
17. Koenderink, J.J.: The structures of images. Biological Cybernetics 50, 363–370 (1984)
18. Babaud, J., Witkin, A.P., Baudin, M., Duda, R.O.: Uniqueness of the Gaussian kernel for scale-space filtering. IEEE Trans. Pattern Anal. Machine Intell. 8(1), 26–33 (1986)
19. Yuille, A.L., Poggio, T.: Scaling theorems for zero-crossings. IEEE Trans. Pattern Anal. Machine Intell. 8(1), 15–25 (1986)
20. Watt, R.J., Morgan, M.J.: A theory of the primitive spatial code in human vision. Vision Research 25, 1661–1674 (1985)
21. Heal, K.M., Hansen, M.L., Richard, K.M.: MapleV Learning Guide. Springer, Heidelberg (1998)
22. Demi, M., Gemignani, V., Paterni, M., Benassi, A.: Real Time Contour Tracking of Cardiovascular Structures with Statistical Filters. In: 5th International Workshop on Nonlinear Signal and Image Processing, pp. 1–5 (2001)
23. Faita, F., Gemignani, V., Giannoni, M., Benassi, A.: A Fully Customizable DSP Based System for Real-Time Imaging. In: Proc. of International Signal Processing Conference - GSPx, pp. 1–5 (2003)
24. Corretti, M.C., Anderson, T.J., Benjamin, E.J., Celermajer, D., Charbonneau, F., Creager, M.A., Deanfield, J., Drexler, H., Gerhard-Herman, M., Herrington, D., Vallance, P., Vita, J., Vogel, R.: Guidelines for the ultrasound assessment of endothelial-dependent flow-mediated vasodilation of the brachial artery: a report of the International Brachial Artery Reactivity Task Force. J. Am. Coll. Cardiol. 39(2), 257–265 (2002)
25. Bots, M.L., Evans, G.W., Riley, W.A., Grobbee, D.E.: Carotid intima-media thickness measurements in intervention studies: design options, progression rates, and sample size considerations: a point of view. Stroke 34(12), 2985–2994 (2003)
26. Wang, J.G., Staessen, J.A., Li, Y., Van Bortel, L.M., Nawrot, T., Fagard, R., Messerli, F.H., Safar, M.: Carotid intima-media thickness and antihypertensive treatment: a meta-analysis of randomized controlled trials. Stroke 37(7), 1933–1940 (2006)
27. De Micheli, E., Caprile, B., Ottonello, P., Torre, V.: Localization and Noise In Edge Detection. IEEE Trans. Pattern Anal. Machine Intell. 11, 1106–1117 (1989)
28. Fleck, M.M.: Some Defects in Finite-Difference Edge Finders. IEEE Trans. Pattern Anal. Machine Intell. 14, 337–345 (1992)

A Unified Mathematical Treatment of Regression Problems in Image Processing

Karlheinz Spindler*

Fachhochschule Wiesbaden, Arbeitsgruppe Mathematik
Kurt-Schumacher-Ring 18, D-65197 Wiesbaden, Germany

Abstract. In this paper we study some optimization problems resulting from image processing tasks in medical applications. These problems are solved using a coordinate-free approach which not only reduces the computational effort in finding the solutions and enhances conceptual clarity, but also leads to closed-form solutions.

Keywords: Object Matching – Automatic Image Analysis – Mathematical Optimization – Medical Applications.

1 Introduction

The increasing availability of pictures in digital form and improved storage and processing capabilities allow the application of more and more sophisticated mathematical methods in pattern recognition and pattern matching and the use of computer vision methods in navigation and control; see [3,4,6,7,9,10] for an overview of the field. Increased mathematical sophistication is not only required to refine algorithms to cope with the growing amounts of data to be processed, but also to generalize the conceptual framework through which application problems are approached.

In this paper we investigate some optimization problems which emerged from the application of image processing techniques to the area of medical diagnostics and computer-aided surgery. It is shown that by using a higher level of mathematical abstraction and working in a coordinate-free manner, it is possible to both enhance conceptual clarity and reduce the required computational effort. Since the emphasis is on conveying an idea rather than tackling a full-scale application in detail, we chose simple and well-known examples, but the approach has the potential of being generalized to highly nontrivial problems which are best formulated as optimization problems on manifolds.

2 Four Sample Problems

The first problem we want to consider stems from attempts to improve diagnostical techniques to detect cervical cancer. The goal is to localize, as precisely

---

* This work was partially supported by the German Federal Ministry of Education and Research (BMBF) under research grant no. 1763X07.

P. Perner and O. Salvetti (Eds.): MDA 2008, LNAI 5108, pp. 108–122, 2008.

as possible, regions in a patient's cervix with tissue anomalies which can be detected in tomographic pictures of the cervical region. To do so, a standard three-dimensional model of a woman's cervical region was developed to which the cervical region of an individual patient is to be matched, allowing for individual variations in body build. This is accomplished by matching some easily identifiable and localizable body features with the corresponding features of the standard model and by then transforming the whole cervical region correspondingly in such a way that collinearity and proportions are preserved. In other words, one looks for an affine transformation which matches given points $P_1, \ldots, P_N$ in space (the patient's body features) with points $Q_1, \ldots, Q_N$ (the corresponding body features of the standard model) as closely as possible. (Similar approaches are used in brain surgery where features of the brain of a specific patient are matched with those listed in a brain atlas such as [12]; cf. [8].) In mathematical terms, this problem can be formulated as follows (cf. [1], [14]).

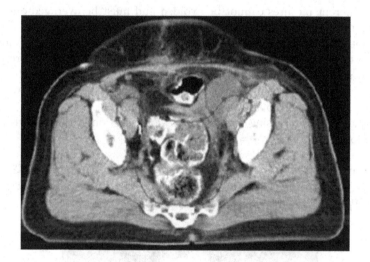

Fig. 1. Cross-section of a woman's cervical region with specified landmarks

(2.1) Affine Pattern Matching Problem. *Given N points $P_i = (x_i, y_i, z_i)$ $\in \mathbb{R}^3$ and N points $Q_i = (u_i, v_i, w_i) \in \mathbb{R}^3$, find an affine transformation $T : \mathbb{R}^3 \to \mathbb{R}^3$ which minimizes the expression $f(T) := \sum_{i=1}^{N} \|T(P_i) - Q_i\|^2$.*

The second problem has its origin in interstitial brachytherapy, which is a medical treatment whereby several biopsy needles supplied with a radioactive source are placed inside a tumor to enable fighting the tumor by radiation from within the body. Apart from the geometrical distribution of the needles inside the tumor tissue, accuracy of positioning of the needles is a critical factor for the success of the therapy. In the setting described here (see [11]), the biopsy needles were equipped with diodes emitting infrared light which was received by a system of four CCD video cameras mounted atop the surgery table; on the other hand, the biopsy needles held markers detectable on tomographic images which were continuously taken for constant supervision. The task of navigation in the positioning of the needles required continuous alignment of the coordinate frames used by the cameras and by the tomographic scanner via a rigid transformation and hence the real-time solution of the following problem (which also occurs in camera calibration; cf. [13]) which sounds similar to (2.1), but requires a completely different solution.

(2.2) Rigid Pattern Matching Problem. *Given N points $P_i = (x_i, y_i, z_i) \in \mathbb{R}^3$ and N points $Q_i = (u_i, v_i, w_i) \in \mathbb{R}^3$, find a rigid transformation $T : \mathbb{R}^3 \to \mathbb{R}^3$ which minimizes the expression $f(T) := \sum_{i=1}^{N} \|T(P_i) - Q_i\|^2$.*

The next problem we want to consider stems from the task of developing a three-dimensional model of a patient's cranium showing features detected on a series of two-dimensional cross-section scans of the cranium obtained by computer tomography. The difficulty arises that errors in locating feature points between different pictures cannot be avoided and must be averaged out during the subsequent image processing. This is done in a two-step approach: first, one tries to determine (using a regression in the total least-squares sense) for each individual cross-section the symmetry line through the detected feature points; second, by superposition of the various cross-sections while matching the axes of symmetry and by then using another regression, one determines the symmetry plane for a three-dimensional model of the cranium.

Thus the image processing requires a two-fold solution of the following problem (one for $d = 2$, one for $d = 3$).

(2.3) Symmetry Hyperplane Problem. *Given N points $P_i \in \mathbb{R}^d$ and N points $\widehat{P}_i \in \mathbb{R}^d$, find a hyperplane reflection $\sigma : \mathbb{R}^d \to \mathbb{R}^d$ such that $f(\sigma) := \sum_{i=1}^{N} \|\sigma(P_i) - \widehat{P}_i\|^2$ becomes minimal.*

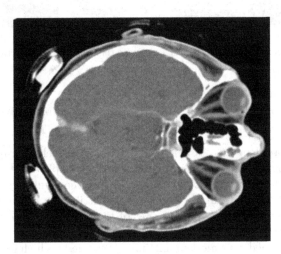

Fig. 2. Image obtained by computer tomography of a patient's head

Remark. Since the reflection at a hyperplane through a point Q with unit normal vector n is given by $\sigma(x) = x - 2\langle x - q, n \rangle n$, we may write the cost function in the form $f(q, n) = \sum_{i=1}^{N} \|p_i - \hat{p}_i - 2\langle p_i - q, n \rangle n\|^2 = \sum_{i=1}^{N} \|p_i - \hat{p}_i\|^2 + 4 \cdot \Phi(q, n)$ where $\Phi(q, n) := \sum_{i=1}^{N} (\langle p_i - \hat{p}_i, n \rangle \langle q - p_i, n \rangle + \langle q - p_i, n \rangle^2)$. (Here we use the convention of identifying a point A with the vector $a = \overrightarrow{OA}$ where O denotes the origin of the chosen coordinate system.)

A similar problem (which can be solved with the same methods) arises when one seeks a hyperplane which fits a given cloud of points as well as possible, where the goodness of fit is measured in the total least-squares sense. (See [15] for a good introduction to total least-squares estimation.) A different approach to symmetry axis detection, using cellular neural networks, is outlined in [2].

(2.4) Regression Hyperplane Problem. *Given N points $P_i \in \mathbb{R}^d$, find a hyperplane $\mathcal{H} \subseteq \mathbb{R}^d$ such that $f(\mathcal{H}) := \sum_{i=1}^{N} \mathrm{dist}(P_i, \mathcal{H})^2$ becomes minimal, where $\mathrm{dist}(P, \mathcal{H})$ denotes the distance from a point P to a hyperplane $\mathcal{H}$.*

Remark. Since $\mathrm{dist}(P, \mathcal{H}) = |\langle p - q, n \rangle|$ where Q is any point in $\mathcal{H}$ and where n is a unit normal vector of $\mathcal{H}$, we may write the cost function in the form $f(q, n) := \sum_{i=1}^{N} \langle p_i - q, n \rangle^2$ where $q, n \in \mathbb{R}^d$ with $\|n\| = 1$.

The functions to be optimized in the above examples have as their domain the set of all affine transformations of $\mathbb{R}^3$, the set of all rigid transformations of $\mathbb{R}^3$, the set of all hyperplane reflections in $\mathbb{R}^d$ and the set of all hyperplanes in $\mathbb{R}^d$ where $d = 2$ or $d = 3$, respectively. A typical approach would be to parametrize these sets by introducing some sort of coordinates and then expressing the function to be minimized as a function of these coordinates. For example, in problem (2.1) we may write an arbitrary affine transformation in the form

$$T(x, y, z) = \begin{bmatrix} a_{11} & a_{12} & a_{13} \\ a_{21} & a_{22} & a_{23} \\ a_{31} & a_{32} & a_{33} \end{bmatrix} \begin{bmatrix} x \\ y \\ z \end{bmatrix} + \begin{bmatrix} b_1 \\ b_2 \\ b_3 \end{bmatrix}$$

and then express the cost functional as a function in the twelve unknowns a_{ij} and b_k where $1 \leq i, j, k \leq 3$, seeking the minimum by equating the partial derivatives with respect to these variables with zero. In problem (2.2) one would either solve the same minimization problem as before, but now under the constraints

$$a_{1i}a_{1j} + a_{2i}a_{2j} + a_{3i}a_{3j} = \delta_{ij} \quad (1 \leq i \leq j \leq 3)$$

which express the fact that $A = (a_{ij})$ is required to be an orthogonal matrix, and then apply the method of Lagrange multipliers; or one would parametrize the matrix A, for example by writing A as

$$\begin{bmatrix} \cos\alpha\cos\beta - \sin\alpha\sin\beta\cos\gamma & \sin\alpha\cos\beta + \cos\alpha\sin\beta\cos\gamma & \sin\beta\sin\gamma \\ -\cos\alpha\sin\beta - \sin\alpha\cos\beta\cos\gamma & -\sin\alpha\sin\beta + \cos\alpha\cos\beta\cos\gamma & \cos\beta\sin\gamma \\ \sin\alpha\sin\gamma & -\cos\alpha\sin\gamma & \cos\gamma \end{bmatrix}$$

in terms of Euler angles, and then treat the problem as an ordinary optimization problem without constraints in the parametrizing variables, treating separately those matrices which are not covered by the Euler angle parametrization and the matrices of determinant -1 (if transformations which do not preserve orientation are allowed in the problem). In problems (2.3) and (2.4) one would write a hyperplane in the form $a_1 x_1 + \cdots + a_d x_d = 1$ or $a_1 x_1 + \cdots + a_d x_d = 0$, depending on whether or not it passes through the origin, and then treat the function to be minimized as a function in the variables $a_1, \ldots, a_d$ under the constraint $a_1^2 + \cdots + a_d^2 = 1$.

Following such a straightforward approach would, in all cases with the exception of (2.1) which is an unconstrained optimization problem, lead to a morass of calculations and is not likely to provide a closed-form solution. On the other hand, by taking a more abstract point of view and working in an essentially coordinate-free way, one not only gains conceptual clarity, but, in fact, arrives at closed-form solutions. (The approach chosen is beneficial in a more general setting; cf. [5].) There is also no need to restrict the investigation to dimensions two and three; all four problems can, without any extra effort, be solved in arbitrary dimensions. The next section gives the concepts and results which are needed for the proposed approach.

3   Notation

We consider a finite-dimensional real vector space equipped with an inner product $\langle \cdot, \cdot \rangle$ and the associated norm $\|v\| := \sqrt{\langle v, v \rangle}$. We do not distinguish in our notation between V as a vector space and V as an affine space; hence given an origin O, a *point* P in V (i.e., an element of the affine space V) will be identified with the *vector* $p = \overrightarrow{OP}$ (i.e., an element of the vector space V). The *barycenter* of a finite set $\{P_1, \ldots, P_N\}$ of points in V is the unique point $\widehat{P}$ with the property that $\widehat{p} = (\sum_{i=1}^{N} p_i)/N$. An *affine transformation* on V is a mapping $T : V \to V$ of the form $Tv = Av + b$ where $A : V \to V$ is a linear mapping and where $b \in V$ is a fixed vector.

The *endomorphism algebra* of V, consisting of all linear mappings ("operators") $A : V \to V$, will be denoted by $\mathrm{End}(V)$. The symbols $\mathbf{0}$ und $\mathbf{1}$ will be used to denote the zero operator and the identity operator, respectively. The *adjoint* of an endomorphism $A : V \to V$ is the unique endomorphism $A^\star : V \to V$ satisfying $\langle A^\star u, v \rangle = \langle u, Av \rangle$ for all $u, v \in V$; if we identify A with a matrix with respect to some orthonormal basis of V, then $A^\star = A^T$ becomes the transpose of A. An operator A is called *self-adjoint* if $A^\star = A$ and *orthogonal* if $A^\star = A^{-1}$. We make the important observation that $\mathrm{End}(V)$ becomes itself an inner product space via the *Frobenius product*

$$\langle\!\langle A, B \rangle\!\rangle := \mathrm{tr}(AB^\star) = \mathrm{tr}(A^\star B) = \mathrm{tr}(BA^\star) = \mathrm{tr}(B^\star A)$$

where tr denotes the trace of an operator. If we identify A and B with real $(n \times n)$-matrices with respect to an orthonormal basis of V, then $\langle\!\langle A, B \rangle\!\rangle = \sum_{i,j=1}^{n} A_{ij} B_{ij}$; i.e., $\langle\!\langle \cdot, \cdot \rangle\!\rangle$ becomes the canonical inner product if we identify the vector space $\mathbb{R}^{n \times n}$ of all real $(n \times n)$-matrices with the vector space $\mathbb{R}^{n^2}$ by writing the entries of an $(n \times n)$-matrix as a column vector of length n^2.

With any two vectors $u, v \in V$ we associate the *tensor product* $u \otimes v \in \text{End}(V)$ which is the linear map $u \otimes v : V \to V$ defined by $(u \otimes v)(x) := \langle x, v \rangle u$ for all $x \in V$; if we identify u and v with elements of $\mathbb{R}^n$ with respect to some orthonormal basis of V then $u \otimes v = uv^T$. The following properties of the tensor product are readily verified:

(1) $u \otimes v$ depends bilinearly on (u, v);
(2) if $v \neq 0$ then the image of $u \otimes v$ is $\mathbb{R}u$;
(3) $(u \otimes v)^\star = v \otimes u$;
(4) $A \circ (u \otimes v) = (Au) \otimes v$ and $(u \otimes v) \circ A = u \otimes (A^\star v)$ for all $A \in \text{End}(V)$;
(5) $(a \otimes b) \circ (c \otimes d) = \langle b, c \rangle \, a \otimes d$;
(6) $\text{tr}(u \otimes v) = \langle u, v \rangle$;
(7) $\langle\!\langle a \otimes b, c \otimes d \rangle\!\rangle = \langle a, c \rangle \langle b, d \rangle$;
(8) $\langle u, Av \rangle = \langle\!\langle u \otimes v, A \rangle\!\rangle$ for all $A \in \text{End}(V)$.

We note that (8) implies that every bilinear form on V can be uniquely represented as a linear form on $\text{End}(V)$, which is the universal property of abstractly defined tensor products. Thus $\text{End}(V)$ with the operation $\otimes$ can be seen as a concrete realization of the abstract tensor product $V \otimes V$, but this is irrelevant for our purposes.

A function $f : V \to \mathbb{R}$ is called *differentiable* at a point $p \in V$ if there is a linear form (necessarily unique) $f'(p) : V \to \mathbb{R}$ such that $f(p + h) = f(p) + f'(p)h + o(\|h\|)$. If this is the case then there is a unique vector $(\nabla f)(p)$ called the *gradient* of f at p such that $f'(p)v = \langle v, (\nabla f)(p) \rangle$ for all $v \in V$. If f is differentiable then, by the Chain Rule, $f'(p)$ can be evaluated by calculating directional derivatives, viz.

$$ f'(p)v \;=\; \frac{\mathrm{d}}{\mathrm{d}t}\bigg|_{t=0} f(p + tv) $$

(or more generally $f'(p)v = (\mathrm{d}/\mathrm{d}t)|_{t=0} f(\alpha(t))$ for any curve $\alpha : (-\varepsilon, \varepsilon) \to V$ with $\alpha(0) = p$ which is differentiable at $t = 0$ and satisfies $\dot{\alpha}(0) = v$). Of course, the gradient of a function at some point depends not just on the vector space V but also on the choice of the inner product. In our setting, the gradient of a differentiable function $f : V \to \mathbb{R}$ at an element $p \in V$ is the unique element $(\nabla f)(p) \in V$ satisfying $(\mathrm{d}/\mathrm{d}t)_{t=0} f(p + tv) = \langle (\nabla f)(p), v \rangle$ for all $v \in V$, and the gradient of a differentiable function $\Phi : \text{End}(V) \to \mathbb{R}$ at $A \in \text{End}(V)$ is the unique element $(\nabla \Phi)(A)$ of $\text{End}(V)$ satisfying $(\mathrm{d}/\mathrm{d}t)|_{t=0} \Phi(A + tB) = \langle\!\langle (\nabla \Phi)(A), B \rangle\!\rangle$ for all $B \in \text{End}(V)$. If differentiability is clear beforehand, the gradient of a function $\Phi : \text{End}(V) \to \mathbb{R}$ can be practically computed by evaluating the expression $(\mathrm{d}/\mathrm{d}t)|_{t=0} \Phi(A + tB)$ and bringing it into the form $\langle\!\langle X, B \rangle\!\rangle$ for some $X \in \text{End}(V)$; this X is then $(\nabla \Phi)(A)$. For a function $f : X \times Y \to \mathbb{R}$ defined on a direct product of two inner-product spaces we denote by $(\nabla_x f)(x_0, y_0)$ the gradient of the function $x \mapsto f(x, y_0)$ at x_0, representing the partial derivative of f with respect to x at (x_0, y_0). We give some examples for the computation of gradients (almost all of which will be used later in solving the problems from section 2).

(3.1) Example. Define $f : V \to \mathbb{R}$ by $f(x) := \langle x - x_0, y \rangle^2$ where $x_0, y \in V$ are fixed vectors. In coordinates, f is a quadratic polynomial in the entries of x; hence it is clear that f is everywhere differentiable. Then

$$
\begin{aligned}
f'(x)v &= \left.\frac{d}{dt}\right|_{t=0} f(x + tv) = \left.\frac{d}{dt}\right|_{t=0} \langle x - x_0 + tv, y \rangle^2 \\
&= \left.\frac{d}{dt}\right|_{t=0} \left(\langle x - x_0, y \rangle^2 + 2t\langle x - x_0, y \rangle\langle v, y \rangle + t^2 \langle v, y \rangle^2 \right) \\
&= 2\langle x - x_0, y \rangle\langle v, y \rangle = \langle 2\langle x - x_0, y \rangle y, v \rangle
\end{aligned}
$$

and hence $(\nabla f)(x) = 2\langle x - x_0, y \rangle y = 2\,(y \otimes y)(x - x_0)$ for all $x \in V$.

(3.2) Example. Define $f : V \to \mathbb{R}$ by $f(x) := \langle x, a \rangle\langle x, b \rangle$ where $a, b \in V$ are fixed vectors. In coordinates, f is a quadratic polynomial in the entries of x; hence it is clear that f is everywhere differentiable. Then

$$
\begin{aligned}
f'(x)v &= \left.\frac{d}{dt}\right|_{t=0} f(x + tv) = \left.\frac{d}{dt}\right|_{t=0} \langle x + tv, a \rangle\langle x + tv, b \rangle \\
&= \left.\frac{d}{dt}\right|_{t=0} \left(\langle x, a \rangle\langle x, b \rangle + t\langle x, a \rangle\langle v, b \rangle + t\langle v, a \rangle\langle x, b \rangle + t^2 \langle v, a \rangle\langle v, b \rangle \right) \\
&= \langle x, a \rangle\langle v, b \rangle + \langle v, a \rangle\langle x, b \rangle = \langle \langle b, x \rangle a + \langle a, x \rangle b, v \rangle
\end{aligned}
$$

and hence $(\nabla f)(x) = \langle b, x \rangle a + \langle a, x \rangle b = (a \otimes b + b \otimes a)(x)$ for all $x \in V$.

(3.3) Example. Define $\Phi : \mathrm{End}(V) \to \mathbb{R}$ by $\Phi(A) := \langle Ap, q \rangle$ where $p, q \in V$ are fixed. Since $\Phi(A)$ depends linearly on the entries of A it is clear that Φ is everywhere differentiable. Then, evaluating derivatives via directional derivatives as explained before, we get

$$
\begin{aligned}
\Phi'(A)B &= \left.\frac{d}{dt}\right|_{t=0} \Phi(A + tB) = \left.\frac{d}{dt}\right|_{t=0} \langle Ap + tBp, q \rangle \\
&= \left.\frac{d}{dt}\right|_{t=0} \left(\langle Ap, q \rangle + t\langle Bp, q \rangle \right) = \langle Bp, q \rangle = \langle q, Bp \rangle = \langle\!\langle q \otimes p, B \rangle\!\rangle
\end{aligned}
$$

where we used property (8) in the last step. Thus $(\nabla \Phi)(A) = q \otimes p$ (independently of A).

(3.4) Example. Define $\Phi : \mathrm{End}(V) \to \mathbb{R}$ by $\Phi(A) := \frac{1}{2}\|Ap - q\|^2$ where $p, q \in V$ are fixed. Being a quadratic polynomial in the entries of A, the function Φ is clearly everywhere differentiable. We compute

$$
\begin{aligned}
\Phi'(A)(B) &= \left.\frac{d}{dt}\right|_{t=0} \Phi(A + tB) = \left.\frac{d}{dt}\right|_{t=0} \frac{1}{2}\|Ap - q + tBp\|^2 \\
&= \left.\frac{d}{dt}\right|_{t=0} \left(\frac{1}{2}\|Ap - q\|^2 + t\langle Ap - q, Bp \rangle + \frac{t^2}{2}\langle Bq, Bq \rangle \right)
\end{aligned}
$$

$$= \langle Ap - q, Bp \rangle = \langle\!\langle (Ap - q) \otimes p, B \rangle\!\rangle$$

where again property (8) was used in the last step; hence $(\nabla\Phi)(A) = (Ap - q) \otimes p$.

(3.5) Example. Define $\Phi : \mathrm{End}(V) \to \mathbb{R}$ by $\Phi(A) := \mathrm{tr}(A)$ where $\mathrm{tr}(A)$ denotes the trace of an endomorphism A. This is a linear functional on $\mathrm{End}(V)$ and as such, of course, differentiable. Since

$$\Phi'(A)B = \left.\frac{\mathrm{d}}{\mathrm{d}t}\right|_{t=0} \mathrm{tr}(A + tB) = \left.\frac{\mathrm{d}}{\mathrm{d}t}\right|_{t=0} (\mathrm{tr}(A) + t\,\mathrm{tr}(B)) = \mathrm{tr}(B) = \langle\!\langle \mathbf{1}, B \rangle\!\rangle$$

for all B, we have $(\nabla\Phi)(A) = \mathbf{1}$ (independently of A).

(3.6) Example. Define $\Phi : \mathrm{End}(V) \to \mathbb{R}$ by $\Phi(A) := \det(A)$. In coordinates, $\det(A)$ is a polynomial in the entries of A; hence Φ is everywhere differentiable. Identifying A with a matrix and denoting by A_i the matrix which is obtained from A by replacing the i-th column by the corresponding column of B, we find, using the linearity of the determinant in each column, that $\det(A + tB) = \det(A) + t \cdot \sum_{j=1}^{n} \det(A_j) +$ higher-order terms in t (with two or more columns of A replaced). Therefore, expanding A_j by the j-th column, we find that

$$\det{}'(A)B = \left.\frac{\mathrm{d}}{\mathrm{d}t}\right|_{t=0} \det(A + tB) = \sum_{j=1}^{n} \det(A_j) = \sum_{j=1}^{n}\sum_{i=1}^{n} (-1)^{i+j} b_{ij} \det(A_{ij})$$

where A_{ij} is obtained from A by striking out the i-th row and the j-th column. Now using the *adjunct* (also called *classical adjoint*) $\Theta := \mathrm{adj}(A)$ of A with entries $\theta_{ij} = (-1)^{i+j} \det(A_{ji})$ for $1 \le i \le n$ this reads

$$\det{}'(A)B = \sum_{j=1}^{n}\sum_{i=1}^{n} b_{ij}\theta_{ji} = \langle\!\langle B, \Theta^T \rangle\!\rangle = \langle\!\langle \mathrm{adj}(A)^T, B \rangle\!\rangle$$

which shows that $(\nabla \det)(A) = \mathrm{adj}(A)^T$. (This observation, in fact, provides a coordinate-free interpretation of the classical adjoint of an operator, namely as the unique operator $\mathrm{adj}(A)$ satisfying $\langle (\nabla \det)(A)u, v \rangle = \langle u, \mathrm{adj}(A)v \rangle$ for all $u, v \in V$.)

4   Solutions to the Sample Optimization Problems

We now show how the problems in Section 2 can be solved using the concepts introduced in Section 3. In all cases it is assumed that V be a finite-dimensional real vector space with an inner product $\langle \cdot, \cdot \rangle$; the notation of Section 3 will be used throughout. We derive necessary conditions for a minimum, but in all cases the existence of a minimum can be easily established by reducing the problem to a situation in which the minimum of a continuous function with compact domain is sought. We start by treating the affine pattern matching problem, which is completely solved by the following result.

(4.1) Theorem. *Let $(P_1, \ldots, P_N)$ and $(Q_1, \ldots, Q_N)$ be two ordered families of points in V. We denote by $\widehat{P}$ and $\widehat{Q}$ the barycenters of the sets $\{P_1, \ldots, P_N\}$ and $\{Q_1, \ldots, Q_N\}$, respectively. If $T : V \to V$ is an affine transformation which minimizes the function $\sum_{i=1}^{N} \|T(P_i) - Q_i\|^2$, then T is given by $Tv = A(v - \widehat{p}) + \widehat{q}$ with a linear map $A : V \to V$ satisfying*

$$A\left(\sum_{i=1}^{N}(p_i - \widehat{p}) \otimes (p_i - \widehat{p})\right) = \sum_{i=1}^{N}(q_i - \widehat{q}) \otimes (p_i - \widehat{p}).$$

Remark. Let $d := \dim(V)$. If amongst the N points P_i there are $d + 1$ in general position (so that $\{P_1, \ldots, P_N\}$ is not contained in a lower-dimensional affine subspace of V) then d of the vectors $p_i - \widehat{p}$ are linearly independent, which implies that $\sum_{i=1}^{N}(p_i - \widehat{p}) \otimes (p_i - \widehat{p})$ is invertible; hence in this case there is a unique solution given by

$$A = \left(\sum_{i=1}^{N}(q_i - \widehat{q}) \otimes (p_i - \widehat{p})\right)\left(\sum_{i=1}^{N}(p_i - \widehat{p}) \otimes (p_i - \widehat{p})\right)^{-1}.$$

Proof. Write $Tv = Av + b$ with a linear mapping $A : V \to V$ and a translation vector b. Since A und b are such that

$$f(A, b) := \sum_{i=1}^{N}\|Ap_i + b - q_i\|^2 = \sum_{i=1}^{N}\left(\|Ap_i - q_i\|^2 + 2\langle Ap_i - q_i, b\rangle + \|b\|^2\right)$$

becomes minimal, the partial derivative with respect to b must vanish, which implies that $0 = (\nabla_b f)(A, b) = 2\sum_{i=1}^{N}(Ap_i - q_i) + 2Nb$ and thus $b = -\sum_{i=1}^{N}(Ap_i - q_i)/N$. Consequently, T is given by

$$Tv = Av - \frac{1}{N}\sum_{i=1}^{N}(Ap_i - q_i) = A\left(v - \frac{1}{N}\sum_{i=1}^{N}p_i\right) + \frac{1}{N}\sum_{i=1}^{N}q_i = A(v - \widehat{p}) + \widehat{q}$$

which yields the (intuitively plausible) result that T must map the barycenter $\widehat{P}$ of the first point set to the barycentre $\widehat{Q}$ of the second point set. By translating the coordinate systems used in both the domain and the range of T, we may assume that each point set has its barycenter in the origin; hence we may assume that $\widehat{p} = \widehat{q} = 0$ and thus have to only consider the cost functional

$$f(A) := \frac{1}{2}\sum_{i=1}^{N}\|Ap_i - q_i\|^2.$$

According to Example (3.4) the gradient of f is given by $(\nabla f)(A) = \sum_{i=1}^{N}(Ap_i - q_i) \otimes p_i$. Since A is optimal, we have $\mathbf{0} = (\nabla f)(A) = \sum_{i=1}^{N}(Ap_i - q_i) \otimes p_i =$

$\sum_{i=1}^{N}(Ap_i) \otimes p_i - \sum_{i=1}^{N} q_i \otimes p_i = A(\sum_{i=1}^{N} p_i \otimes p_i) - \sum_{i=1}^{N} q_i \otimes p_i$. This is the claim. ∎

The following theorem finds the optimal matching transformation under the condition that only rigid transformations are permitted. Again, a solution can be found which is essentially in closed form.

(4.2) Theorem. *Let* $(P_1, \ldots, P_N)$ *and* $(Q_1, \ldots, Q_N)$ *be two ordered families of points in* V. *We denote by* $\widehat{P}$ *and* $\widehat{Q}$ *the barycenters of the sets* $\{P_1, \ldots, P_N\}$ *and* $\{Q_1, \ldots, Q_N\}$, *respectively. If* $T : V \to V$ *is a rigid motion which minimizes* $\sum_{i=1}^{N} \|T(P_i) - Q_i\|^2$, *then* T *is given by* $Tv = A(v - \widehat{p}) + \widehat{q}$ *where* $A^\star A = 1$ *and* ΘA *is self-adjoint with*

$$\Theta := \sum_{i=1}^{N}(p_i - \widehat{p}) \otimes (q_i - \widehat{q}).$$

Remark. In matrix terms the optimality condition means that ΘA is a symmetric matrix. This condition can be expressed as a set of $(n^2 - n)/2$ equations (linear in the entries of A) for the off-diagonal elements of ΘA from which A (which is an element of the $(n^2 - n)/2$-dimensional orthogonal group of V) can be determined.

Proof. We can show as in (4.1) that any optimal T must map $\widehat{P}$ to $\widehat{Q}$. Thus we can assume that $\widehat{p} = 0$ and $\widehat{q} = 0$ and are thus asked to find an orthogonal transformation A such that $\sum_{i=1}^{N} \|Ap_i - q_i\|^2$ becomes minimal. Since $\|Ap_i - q_i\|^2 = \|Ap_i\|^2 - 2\langle Ap_i, q_i \rangle + \|q_i\|^2 = \|p_i\|^2 - 2\langle Ap_i, q_i \rangle + \|q_i\|^2$, this is tantamount to maximizing the function $\Phi : \text{End}(V) \to \mathbb{R}$ given by

$$\Phi(A) := \sum_{i=1}^{N} \langle Ap_i, q_i \rangle$$

under the constraint $A^\star A = 1$; note that $(\nabla \Phi)(A) = \sum_{i=1}^{N} q_i \otimes p_i$ according to example (3.3). Now the operator-valued constraint $A^\star A = 1$ is equivalent to the family of all scalar-valued constraints $\Phi_{u,v}(A) = 0$ where $\Phi_{u,v}(A) := \langle u, A^\star Av \rangle - \langle u, v \rangle = \langle Au, Av \rangle - \langle u, v \rangle$ for $u, v \in V$. We compute

$$\Phi'_{u,v}(A)B = \frac{d}{dt}\Big|_{t=0} \Phi_{u,v}(A + tB) = \frac{d}{dt}\Big|_{t=0} \left(\langle Au + tBu, Av + tBv \rangle - \langle u, v \rangle \right)$$

$$= \frac{d}{dt}\Big|_{t=0} \left(\langle Au, Av \rangle - \langle u, v \rangle + t \left(\langle Bu, Av \rangle + \langle Au, Bv \rangle \right) + t^2 \langle Bu, Bv \rangle \right)$$

and hence

$$\Phi'_{u,v}(A)(B) = \langle Bu, Av \rangle + \langle Au, Bv \rangle = \langle Av, Bu \rangle + \langle Au, Bv \rangle$$
$$= \langle\!\langle (Av) \otimes u, B \rangle\!\rangle + \langle\!\langle (Au) \otimes v, B \rangle\!\rangle = \langle\!\langle A \circ (v \otimes u), B \rangle\!\rangle + \langle\!\langle A \circ (u \otimes v), B \rangle\!\rangle$$
$$= \langle\!\langle A \circ (v \otimes u) + A \circ (u \otimes v), B \rangle\!\rangle = \langle\!\langle A \circ (u \otimes v + v \otimes u), B \rangle\!\rangle.$$

This shows that $(\nabla\Phi_{u,v})(A) = A(u \otimes v + v \otimes u)$. If A is optimal then, by the method of Lagrange multipliers, we can write $(\nabla\Phi)(A) = \sum_{i=1}^{N} q_i \otimes p_i = \Theta^\star$ as a linear combination of the gradients $(\nabla\Phi_{u,v})(A) = A(u \otimes v + v \otimes u)$ where $u, v \in V$ are arbitrary. Since the linear combinations of endomorphisms of the form $u \otimes v + v \otimes u$ are exactly the self-adjoint endomorphisms, this means that $\Theta^\star = AS$ with a self-adjoint operator S, i.e., that $A^{-1}\Theta^\star = A^\star\Theta^\star = (\Theta A)^\star$ is self-adjoint, i.e., that ΘA is self-adjoint. ∎

To determine A explicitly we observe that $\Theta^\star = AS$ implies that $\Theta\Theta^\star = S^\star A^\star AS = S^\star S = S^2$ so that S is a square root of $\Theta\Theta^\star$. Now A maximizes

$$\Phi(A) = \sum_{i=1}^{N} \langle Ap_i, q_i \rangle = \sum_{i=1}^{N} \mathrm{tr}\big((Ap_i) \otimes q_i\big) = \sum_{i=1}^{N} \mathrm{tr}\big(A(p_i \otimes q_i)\big)$$

$$= \mathrm{tr}\Big(\sum_{i=1}^{N} A(p_i \otimes q_i)\Big) = \mathrm{tr}\Big(A\big(\sum_{i=1}^{N} p_i \otimes q_i\big)\Big) = \mathrm{tr}(A\Theta)$$

amongst all A such that $A^\star A = \mathbf{1}$ (or, depending on the problem formulation, all A such that $A^\star A = \mathbf{1}$ and $\det(A) = 1$). Writing $C = \Theta^\star$, the task is to maximize $\mathrm{tr}(C^\star A)$. The solution to this task is given by the following result.

(4.3) Theorem. *Let C be an endomorphism of a finite-dimensional real vector space V with singular values $\sigma_1 \geq \cdots \geq \sigma_n \geq 0$, and let $A \in \mathrm{SO}(V)$ or $A \in \mathrm{O}(V)$ be such that $\Phi(A) := \mathrm{tr}(C^\star A)$ becomes maximal.*

(a) If C has full rank and if the optimization is performed over the full orthogonal group $\mathrm{O}(V)$ then this problem has a unique solution, namely, $A = CS^{-1}$ where $S := \sqrt{C^\star C}$ is the unique positive definite square root of $C^\star C$. If $(e_1, \ldots, e_n)$ is any orthonormal basis of eigenvectors for the eigenvalues $\sigma_1, \ldots, \sigma_n$ of S, we have $S = \sum_{i=1}^{n} \sigma_i e_i \otimes e_i$, yielding the maximum value $\Phi(A) = \sum_{i=1}^{n} \sigma_i$.

(b) If C has full rank and if the optimization is performed over the special orthogonal group $\mathrm{SO}(V)$, then, given any orthonormal basis $(e_1, \ldots, e_n)$ of eigenvectors of S, a solution (unique if and only if the singular value σ_n has multiplicity one) is $A = CS^{-1}$ where $S := \sum_{i=1}^{n-1} \sigma_i e_i \otimes e_i + (\det C)\sigma_n e_n \otimes e_n$, yielding the maximum value $\Phi(A) = \sum_{i=1}^{n-1} \sigma_i + (\det C)\sigma_n$.

(c) If C does not have full rank, then an element $A \in \mathrm{SO}(V)$ or $A \in \mathrm{O}(V)$ yields the maximum value $\Phi(A) = \sum_{i=1}^{n} \sigma_i = \sum_{i=1}^{r} \sigma_i$ (where r is the rank of C) if and only if A has the form

$$A = W^\star \begin{bmatrix} \mathbf{1}_r & 0 \\ 0 & \widehat{B} \end{bmatrix} V$$

where $C = W^\star DV$ is a singular value decomposition of C and where $\widehat{B}$ is an orthogonal operator in dimension $\dim(V) - r$.

Proof. The compactness of $\mathrm{SO}(V)$ and $\mathrm{O}(V)$ guarantees the existence of a maximum, and we know already that if A is optimal then $C = AS$ for some square root of $C^\star C$, i.e., some S such that $S^\star = S$ and $S^2 = C^\star C$. Now if C has full rank then any such S is invertible, and $A = CS^{-1}$ is orthogonal for any such S, whence $\Phi(A) = \operatorname{tr}(C^\star A) = \operatorname{tr}(C^\star CS^{-1}) = \operatorname{tr}(S^2 S^{-1}) = \operatorname{tr}(S) = \sum_{i=1}^n \lambda_i$ where $\lambda_1, \ldots, \lambda_n$ are the eigenvalues of S. Since $S^2 = C^\star C$, we may assume that the λ_i are ordered in such a way that $\lambda_i = \pm\sigma_i$. Now $\operatorname{tr}(S) = \sum_{i=1}^n \lambda_i$ becomes maximal if and only if $\lambda_i = +\sigma_i$ for all i, so that $S = \sum_{i=1}^n \sigma_i\, e_i \otimes e_i = \sqrt{C^\star C}$ where $(e_1, \ldots, e_n)$ is any orthonormal basis of associated eigenvectors. This yields the sought maximum if the domain of maximization is all of $\mathrm{O}(V)$. If the optimization is performed over all $A \in \mathrm{SO}(V)$ only and if the above maximum is attained in $\mathrm{O}(V) \setminus \mathrm{SO}(V)$, then the signs of an odd number of eigenvalues must be changed while keeping the sum of these eigenvectors as large as possible. This means changing only the sign of the smallest singular value of C. Thus if the optimization is performed over $\mathrm{SO}(V)$, the solution is given by $S = \sum_{i=1}^{n-1} \sigma_i\, e_i \otimes e_i + \operatorname{sign}(\det C)\sigma_n\, e_n \otimes e_n$.

If C does not have full rank, we can use a singular value decomposition $C = W^\star DV$ and see that $\Phi(A) = \operatorname{tr}(C^\star A) = \operatorname{tr}(V^\star D^\star WA) = \operatorname{tr}(D^\star WAV^\star)$ is maximized by A if and only if $\psi(B) := \operatorname{tr}(D^\star B)$ is maximized by $B := WAV^\star$. From the above, any optimal B is such that $D = BT$ where $T^\star = T$ and $T^2 = D^\star D = D^2$. Now the equation $D = BT$ takes the block form

$$\begin{bmatrix} D_1 & 0 \\ 0 & 0 \end{bmatrix} = \begin{bmatrix} B_1 & B_2 \\ B_3 & B_4 \end{bmatrix} \begin{bmatrix} T_1 & 0 \\ 0 & 0 \end{bmatrix} = \begin{bmatrix} B_1 T_1 & 0 \\ B_3 T_1 & 0 \end{bmatrix}$$

where $D_1 = \operatorname{diag}(\sigma_1, \ldots, \sigma_r)$ with the nonzero singular values $\sigma_1 \geq \cdots \geq \sigma_r > 0$ of C. This immediately implies $B_3 = 0$ and then, since B is orthogonal, also $B_2 = 0$. But then $\operatorname{tr}(D^\star B) = \operatorname{tr}(D_1^\star B_1)$ becomes maximal if and only if $B_1 = D_1 S_1^{-1} = \mathbf{1}$ where $S_1 = D_1$ is the unique positive definite square root of D_1^2 and where B_4 is arbitrary. Since $B_4 =: \widehat{B}$ can always be chosen in such a way that $\det(A) = 1$, this yields the solution.   ∎

Remark. The decision which of the possible square-roots of $C^\star C$ yields the sought maximum can also be made by invoking the second-derivative criterion. Namely, if A maximizes $\Phi(A) := \operatorname{tr}(C^\star A)$, then, for any given skew-symmetric U, the function $\varphi(t) := \operatorname{tr}\big(C^\star A \exp(tU)\big) = \operatorname{tr}\big(S^\star A^\star A \exp(tU)\big) = \operatorname{tr}\big(S \exp(tU)\big)$ takes a maximum at $t = 0$, which implies that $0 \geq \ddot\varphi(0) = \operatorname{tr}(SU^2)$. If $a, b \in V$ are any given vectors, we can apply this condition with $U := a \otimes b - b \otimes a$ and hence $U^2 = \langle a, b \rangle(a \otimes b + b \otimes a) - \|a\|^2(b \otimes b) - \|b\|^2(a \otimes a)$. Choosing a spectral decomposition $S = \sum_{i=1}^n \lambda_i e_i \otimes e_i$ where $(e_1, \ldots, e_n)$ is an orthonormal basis of V such that e_i is an eigenvector of S associated with the eigenvalue λ_i and writing $a = \sum_{i=1}^n a_i e_i$ and $b = \sum_{i=1}^n b_i e_i$, we find that $(e_i \otimes e_i)U^2$ equals

$$\langle a, b \rangle\big(\langle a, e_i \rangle e_i \otimes b + \langle b, e_i \rangle e_i \otimes a\big) - \|a\|^2 \langle b, e_i \rangle(e_i \otimes b) - \|b\|^2 \langle a, e_i \rangle(e_i \otimes a)$$

and hence that $0 \geq \operatorname{tr}(SU^2) = \sum_{i=1}^n \lambda_i \operatorname{tr}\big((e_i \otimes e_i)U^2\big) = \sum_{i=1}^n \lambda_i\big(2\langle a, b \rangle a_i b_i - \|a\|^2 b_i^2 - \|b\|^2 a_i^2\big) = -\sum_{i=1}^n \lambda_i \|b_i a - a_i b\|^2$. Letting $a := e_k$ and $b := e_\ell$, this becomes

$$0 \leq \sum_{i=1}^n \lambda_i \|b_i e_k - a_i e_\ell\|^2 = \sum_{i=1}^n \lambda_i(\delta_{ik}^2 + \delta_{i\ell}^2) = \lambda_k + \lambda_\ell.$$

Hence $\lambda_k + \lambda_\ell \geq 0$ whenever $k \neq \ell$. This leaves only two possibilities: either $\lambda_i \geq 0$ for $1 \leq i \leq n$ (in which case S is positive semidefinite), or else $n-1$ of the eigenvalues of S are strictly positive and one is strictly negative, with absolute value smaller than any of the others (in which case S is automatically invertible). Now we can apply the same kind of reasoning as in the proof given before. ∎

We conclude by solving the two remaining regression problems described in Section 2. The symmetry hyperplane problem is completely solved by Theorem (4.4), the regression hyperplane problem by Theorem (4.5).

(4.4) Theorem. *Given N points P_i and N points $\widehat{P}_i$ in V, let $\mathcal{H}$ be the hyperplane such that the reflection σ at $\mathcal{H}$ minimizes the expression $\sum_{i=1}^{N} \|\widehat{P}_i - \sigma(P_i)\|^2$. Then $\mathcal{H}$ passes through the barycenter S of the point set $\{P_1, \ldots, P_N, \widehat{P}_1, \ldots, \widehat{P}_N\}$, and each of its normal vectors is an eigenvector (with respect to the smallest eigenvalue) of*

$$\sum_{i=1}^{N} \left((p_i - s) \otimes (\widehat{p}_i - s) + (\widehat{p}_i - s) \otimes (p_i - s) \right).$$

Proof. Writing the hyperplane in the form $\{x \in V \mid \langle x - q, n \rangle = 0\}$ with $\|n\| = 1$, it is clear from the problem formulation in (2.3) that the goal is to minimize the function $\Phi(q, n) := \sum_{i=1}^{N} \left(\langle p_i - \widehat{p}_i, n \rangle \langle q - p_i, n \rangle + \langle q - p_i, n \rangle^2 \right)$. The gradient with respect to q is

$$(\nabla_q \Phi)(q, n) = \sum_{i=1}^{N} \left(\langle p_i - \widehat{p}_i, n \rangle n + 2 \langle q - p_i, n \rangle n \right) = \sum_{i=1}^{N} \langle 2q - p_i - \widehat{p}_i, n \rangle n;$$

thus the condition $(\nabla_q \Phi)(q, n) = 0$, necessary for optimality, yields $\langle q - s, n \rangle = 0$ where $s := \sum_{i=1}^{N}(p_i + \widehat{p}_i)/(2N)$; this shows that the barycenter S of the set of all points P_i and $\widehat{P}_i$ where $1 \leq i \leq N$ lies in the optimal hyperplane we are looking for. Thus, after selecting this barycenter as origin, we may assume that $q = 0$ so that the cost function takes the form

$$\Phi(n) := \sum_{i=1}^{N} \left(\langle p_i - \widehat{p}_i, n \rangle \langle -p_i, n \rangle + \langle p_i, n \rangle^2 \right) = \sum_{i=1}^{N} \langle p_i, n \rangle \langle \widehat{p}_i, n \rangle.$$

This function has to be minimized under the constraint $G(n) = 0$ where $G(n) := \|n\|^2 - 1$; thus there is a Lagrange multiplier λ such that $(\nabla \Phi)(n) = \lambda \cdot (\nabla G)(n)$ which, using Example (3.2), means that

$$\sum_{i=1}^{N}(p_i \otimes \widehat{p}_i + \widehat{p}_i \otimes p_i)n = 2\lambda n \qquad (1)$$

which shows that n is an eigenvector of $\sum_{i=1}^{N}(p_i \otimes \widehat{p}_i + \widehat{p}_i \otimes p_i)$. Moreover, (1) implies that $2\lambda = \langle 2\lambda n, n \rangle = 2\sum_{i=1}^{N}\langle p_i, n \rangle \langle \widehat{p}_i, n \rangle = \Phi(n)$ is just the value of the cost function; hence the smallest eigenvalue must be chosen to obtain the minimal cost. ∎

(4.5) Theorem. *Given N points P_i in V, let $\mathcal{H}$ be an affine hyperplane in V which miminizes the expression $\sum_{i=1}^{N} \mathrm{dist}(P_i, \mathcal{H})^2$. Then $\mathcal{H}$ passes through the barycenter $\widehat{P}$ of the set $\{P_1, \ldots, P_N\}$, and each of its normal vectors is an eigenvector of*

$$\sum_{i=1}^{N} (p_i - \widehat{p}) \otimes (p_i - \widehat{p})$$

with respect to the smallest eigenvalue. (Note that $\sum_{i=1}^{N} (p_i - \widehat{p}) \otimes (p_i - \widehat{p})$ is positive semi-definite so that all the eigenvalues are nonnegative.)

Proof. Assume that Q is an arbitrary point in $\mathcal{H}$ and that n is a unit normal vector of $\mathcal{H}$; then $\mathrm{dist}(P_i, H) = |\langle p_i - q, n \rangle|$ for $1 \leq i \leq N$. Hence the cost function is $F(q, n) := \sum_{i=1}^{N} \langle q - p_i, n \rangle^2$, which is to be minimized on $V \times \mathbb{S}$ where $\mathbb{S}$ is the unit sphere in V. Optimality requires $0 = (\nabla_q F)(q, n) = \sum_{i=1}^{N} 2\langle p_i - q, n \rangle n$ where we used (3.1) in the last equation; consequently, $0 = \sum_{i=1}^{N} \langle p_i - q, n \rangle = \sum_{i=1}^{N} \langle p_i, n \rangle - N \cdot \langle q, n \rangle = \langle \sum_{i=1}^{N} p_i, n \rangle - N \cdot \langle q, n \rangle = N \cdot \langle \widehat{p}, n \rangle - N \cdot \langle q, n \rangle$ so that $\langle \widehat{p} - q, n \rangle = 0$, which implies that $\widehat{P} \in \mathcal{H}$. Thus we arrive at the (intuitively plausible) result that any optimizing hyperplane must pass through the barycenter of the point cloud considered. Translating any coordinate system through this barycenter, we may thus assume that $\widehat{p} = 0$; then $\mathcal{H}$ takes the form $\mathcal{H} = \{x \in V \mid \langle x, n \rangle = 0\}$ (where n still needs to be determined). Thus we have to minimize the cost functional $F(n) := \sum_{i=1}^{N} \langle p_i, n \rangle^2$ under the constraint $G(n) = 0$ where $G(n) := \|n\|^2 - 1$. If n is optimal in this situation then there is a Lagrange multiplier λ such that $(\nabla F)(n) = \lambda \cdot (\nabla G)(n)$ which, according to (3.1), means $(\sum_{i=1}^{N} (p_i \otimes p_i))n = \lambda n$. This shows that n is an eigenvector of $\sum_{i=1}^{N} p_i \otimes p_i$. Moreover, from $\sum_{i=1}^{N} (p_i \otimes p_i)n = \lambda n$ we find that $\lambda = \langle \lambda n, n \rangle = \langle \sum_{i=1}^{N} (p_i \otimes p_i)n, n \rangle = \sum_{i=1}^{N} \langle p_i, n \rangle^2 = F(n)$ is just the value of the cost function; hence the smallest eigenvalue must be chosen to obtain the minimal cost. ∎

(4.6) Special Case. *If in the situation of (4.5) all points P_i lie on a line g then there is a unique optimizing hyperplane, namely the plane passing through the barycenter of the point set $\{P_1, \ldots, P_N\}$ which is perpendicular to g.*

Proof. Let v be a vector spanning g; then each point P_i can be written as $p_i = \widehat{p} + \lambda_i v$ with suitable coefficients $\lambda_i \in \mathbb{R}$. Then, as shown in (4.5), every normal vector n of an optimizing hyperplane is an eigenvector of $\sum_{i=1}^{N} (p_i - \widehat{p}) \otimes (p_i - \widehat{p}) = (\sum_{i=1}^{N} \lambda_i^2) v \otimes v$ and hence an eigenvector of $v \otimes v$. But the only eigenvector (up to a scalar factor) of $v \otimes v$ is v; whence the claim. ∎

Acknowledgments. The support of Prof. Dr. Detlef Richter (Fachhochschule Wiesbaden) and Dr. Gerd Straßmann (Universitätsklinikum Marburg) in introducing me to the problems described in this article and providing figures 1 and 2 is gratefully acknowledged.

References

1. Arun, K.S., Huang, T.S., Blostein, S.D.: Least-Squares Fitting of Two 3-D Point Sets. IEEE Transactions on Pattern Analysis and Machine Intelligence 9(5), 698–700 (1987)
2. Costantini, G., Casali, D.: Detection of Symmetry Axis by a CNN-based Algorithm. In: Proc. 11th WSEAS Conf. Circuits, pp. 46–49. WSEAS Press, Agios Nikolaos (2007)
3. Faugeras, O.: Three-Dimensional Computer Vision: A Geometric Viewpoint. MIT Press, Cambrige (1993)
4. Hartley, R.I., Zisserman, A.W.: Multiple View Geometry in Computer Vision. Cambridge University Press, Cambridge (2001)
5. Hüper, K., Helmke, U.: Jacobi-type Methods in Computer Vision: A Case Study. Zeitschrift für Angewandte Mathematik und Mechanik 78(suppl. 3), 945–948 (1998)
6. Kanatani, K.: Geometric Computation for Machine Vision. Oxford University Press, Oxford (1993)
7. Ma, Y., Soatto, S., Kosečkà, J., Sastry, S.S.: An Invitation to 3-D Vision. Springer, New York (2004)
8. Nowinski, W.L., Yang, G.L., Yeo, T.T.: Computer-Aided Stereotactic Functional Neurosurgery Enhanced by the Use of the Multiple Brain Atlas Database. IEEE Transactions on Medical Imaging 19(1), 62–69 (2000)
9. Olver, P.J., Tannenbaum, A. (eds.): Mathematical Methods in Computer Vision. Springer, New York (2003)
10. Picci, G., Gilliam, D.S. (eds.): Dynamical Systems, Control, Coding, Computer Vision. Birkhäuser, Basel, Boston, Berlin (1999)
11. Richter, D., La Torre, F., Egger, J., Straßmann, G.: Tetraoptical Camera System for Medical Navigation. In: Proc. 17th Biennial International Eurasip Conference Biosignal, Brno (2004)
12. Talairach, J., Tournoux, P.: Co-Planar Stereotaxic Atlas of the Human Brain. Thieme-Verlag, Stuttgart (1988)
13. Tsai, R.Y.: A Versatile Camera Calibration Technique for High Accuracy 3D Machine Vision Metrology Using Off-the-Shelf TV Cameras and Lenses. IEEE Journal of Robotics and Automation RA-3(4), 323–344 (1987)
14. Umeyama, S.: Least-Squares Estimation of Transformation Parameters Between Two Point Patterns. IEEE Transactions on Pattern Analysis and Machine Intelligence 13(4), 376–380 (1991)
15. van Huffel, S., Vandewalle, J.: The Total Least Squares Problem. Society for Industrial and Applied Mathematics (SIAM), Philadelphia (PA) (1991)

Multi-scale Representation and Persistency for Shape Description

Davide Moroni[1], Mario Salvetti[2], and Ovidio Salvetti[1]

[1] Institute of Information Science and Technologies (ISTI),
Italian National Research Council (CNR), Pisa, Italy
{davide.moroni,ovidio.salvetti}@isti.cnr.it
[2] Department of Mathematics, University of Pisa, Pisa, Italy
salvetti@dm.unipi.it

Abstract. Extraction, organization and exploitation of topological features are emerging topics in computer vision and graphics. However, such kind of features often exhibits weak robustness with respect to small perturbations and it is often unclear how to distinguish truly topological features from topological noise. In this paper, we present an introduction to persistence theory, which aims at analyzing multi-scale representations from a topological point of view. Besides, we extend the ideas of persistency to a more general setting by defining a set of discrete invariants.

1 Introduction

Extraction, organization and exploitation of topological features are emerging topics in computer vision and graphics. However, such kind of features is often very sensitive to small perturbations in the object to be studied. For this reason, in classical approaches, *ad hoc* cleaning of the datasets has been used to achieve stability in the computation of topological features. However, topological noise is not always easily distinguishable from non-spurious topological features; thus analyzing datasets via a multiscale procedure is tempting.

Persistency theory is a quite recent theory (see e.g. [1]) which offers the possibility to build multi-scale hierarchical representations for objects related to computer vision, and to analyze and track their topological features. In particular, persistency may describe at which scale a topological feature (e.g. a hole) is created and when it is annihilated (e.g. when the hole has been filled) in a multi-scale representation. Topological features having long lives are more robust and, likely, more salient. In addition, persistency analysis may be applied not only to the original objects, but also to derived spaces in which geometrical properties are coded in a more explicit way. For example in [2] the original object is replaced by its *tangent complex*; a multi-scale representation of the tangent complex is then obtained by considering a filtration based on the curvature. Thus, a blending of geometric information and topology is achievable by persistency analysis. The synergy between the description power of geometry and the discriminative power of algebraic topology invariants is appealing for shape characterization and retrieval applications. In [3], persistency is applied to point cloud data curves in order to extract the

P. Perner and O. Salvetti (Eds.): MDA 2008, LNAI 5108, pp. 123–138, 2008.
© Springer-Verlag Berlin Heidelberg 2008

barcode invariant. A similarity measures for barcodes is then used for shape classification. In [4,5], persistency of 0-dimensional homology (under the original name of size theory [6]) is applied to the skeletal graph of 3D models. The freedom in the choice of the function driving skeletal extraction coupled with the filtrations arising from a wide-ranging collection of shape descriptors allows the construction of suitable similarity measures for 3D models.

Further, persistence has found applications also in biochemistry, in the 3D modeling of proteins and their contact interactions (see e.g. [7,8]).

In Section 2, we motivate and provide an introduction to persistency theory. With respect to [1], which is mainly intended for for computational geometers and combinatorialists, this survey section should be readable with a minimum of background.

In the second part of this paper (Section 3), which is of research nature, we treat multi-dimensional persistency. In particular, after introducing a new filtration of the medial axis -which can capture geometrical features like sharp corner or the presence of parallel segments in the boundary of an object- we attempt to fuse, in the framework of multidimensional persistence, the information extracted from this filtration with an already-known filtration of the medial axis [9].

Our work on multidimensional persistency, although being similar in spirit to [10], has the merit to be more general, in the sense that we are able to treat filtrations organized over an arbitrary partially ordered set. In this more general setting, we are still able to detect the time of creation of topological features and to encode it in a discrete invariant. The lifetime of a topological feature is also rigourously defined.

2 Persistency

2.1 The Basic Idea

To illustrate the basic idea underlying the theory of persistency, we consider a toy example related to digital terrain elevation, namely the analysis of the one-dimensional relief shown in Figure 1, to which we will refer throughout this discussion. The relief may be represented formally as the graph of the function $h : \mathbb{R} \to \mathbb{R}$ where h is the height of the relief. From a heuristic point of view, we would like to assign less importance to the small bump on the left side of the graph with respect to the strong and isolated peak on the right side. To single out these properties, consider for $\lambda \in \mathbb{R}$ the sublevel set $\Gamma_\lambda = h^{-1}(-\infty, \lambda]$. It is convenient to analyze how the topology of Γ_λ changes as we increase λ. For λ less than the minimum height λ_1 of the relief, Γ_λ is empty. When λ exceeds λ_1, the sublevel set Γ_λ consists in a single connected component P_1. A second new connected component P_2 is created when λ exceeds λ_2 and a third one P_3 when λ exceeds λ_3. The two youngest connected components are merged when λ exceeds the local maximum λ_4. Finally, for $\lambda > \lambda_5$ all the connected components are merged. In particular, notice that shape transitions occur at local maxima and minima of the height function; namely exceeding a minimum creates a new connected component, while exceeding a maximum merges two connected components. Using λ as a time parameter, it is possible to keep track of the "life" of a connected component in a way that is compatible with the previous heuristic criterion. First of all, we may say that the connected component P_1 has birth time λ_1 and similarly for P_2 and P_3. When a local maximum

λ_{Max} is exceeded, we may define as a rule that the youngest connected component (between the two connected components that are merged there) dies. By definition, it's death time is λ_{Max}. Thus P_1 is created at λ_1 and never dies. P_2 is created at λ_2 and dies at λ_5. By contrast P_3 is created at λ_3 and dies immediately after at λ_4. In particular, the small bump on the left side is identified as a scarcely persistent feature, quantified by the short lifetime $\lambda_4 - \lambda_3$. An outlook table for persistency is obtained by mapping each connected component to a point in the $2-$dimensional space whose coordinates correspond respectively to the birth and death time. Then the vertical distance from the diagonal represents the lifetime of the topological features (see Figure 1, right side).

To recap, we have analyzed and discovered the interdependency and persistency of topological features of the relief by introducing a family (parameterized by λ) of nested topological spaces (the sublevel sets Γ_λ) and *tracking* topological features among the members of the family. Basically, this is the philosophy underlying persistency; we will show in the section below that nested families of topological spaces arise naturally in the study of objects related to computer vision and graphics, while in Section 2.3 we will see that more refined and general algebraic topology tools are available for "feature tracking".

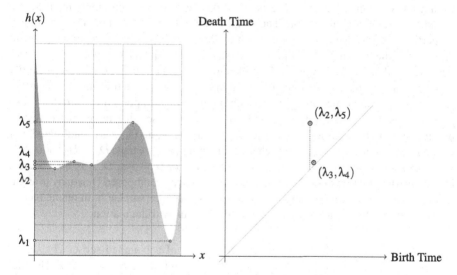

Fig. 1. On the left side a relief considered as the graph of a single variable function; on the right side persistence diagram of the relief: the coordinates are respectively the birth and death time, circles correspond to topological changes that have been matched during the persistency analysis. The vertical distance from the diagonal represents lifetime.

2.2   Examples of 1-Parameter Filtrations

Point Cloud Data and α-Shapes. Often the raw data to be analyzed by pattern recognition methods –or to be explored by visualization techniques– consist in point cloud data, i.e. a finite collection C of points in the n-dimensional real space $\mathbb{R}^n$ without any further geometrical and topological structure. This is also the case of the raw data provided by some 3D-scanner, commonly used for three-dimensional object reconstruction (see e.g. [11]). The concept of α-shapes ([12]; see also [13,14]) provides a method to study at various scales topological properties of point cloud data. The underlying geometric construction can be briefly described as follows. For a point $x \in \mathbb{R}^n$, let $\bar{B}_{x,\alpha}$ be the closed ball of radius α centered in x. The point cloud data C may be α-*thickened* by replacing it with a collection of closed balls:

$$C_\alpha := \{\bar{B}_{x,\alpha} \,|\, x \in C\}$$

We may then partition the union of the balls in C_α into Voronoi regions: for a ball $\bar{B}_{x,\alpha}$, its Voronoi region $V_{x,\alpha}$ consists in the points which are closer to $\bar{B}_{x,\alpha}$ with respect to the other balls in the collection C_α. The intersections among the various Voronoi regions give rise to a rich combinatorics that we may study by defining the so-called *dual complex* D_α. This is a simplicial complex (that is roughly speaking a union of points, vertices, triangles and so on satisfying some precise mathematical properties) defined as follows (see Figure 2). If the regions $V_{x_1,\alpha}$ and $V_{x_2,\alpha}$ intersect, we join x_1 and x_2 with a segment. If three regions $V_{x_1,\alpha}, V_{x_2,\alpha}, V_{x_3,\alpha}$ have non-empty intersection, we add to D_α a triangle having x_1, x_2, x_3 as vertices. If four regions have non-empty intersection, we add a tetrahedron, and so on. The simplicial complex D_α captures the topological properties of C_α; indeed, it is not difficult to show that D_α is a deformation retract of the union of the balls in C_α and thus the two spaces have equivalent topological description. Notice that we have converted a collection of points into a one-parameter family of simplicial complexes, a more rich combinatorial and geometric object. Nevertheless, the members of the family come with an additional important feature that make possible to study them wholesome. Indeed it is easy to see that increasing α results possibly only in the addition of faces to D_α. In particular, no faces are annihilated, nor the geometric realization of them is changed. In summary, we have $D_\alpha \subset D_\beta$ for $\alpha \leq \beta$. In the extreme case $\alpha = 0$, we have $D_0 = C$, while for $\alpha = +\infty$, $D_{+\infty}$ is the classical Delaunay triangulation of C [15]. In particular, we have decomposed Delaunay triangulation by providing an increasing family of spaces that approximate it; in mathematical terminology, α-shapes define a *filtration* of the Delaunay triangulation.

The λ-Medial Axis. The classical *medial axis* $\mathcal{M}$ of a bounded subset $\Omega \subset \mathbb{R}^n$ is the set of points $x \in \Omega$ having at least two closest point in the boundary $\partial\Omega$. The medial axis captures all the topological properties of Ω as described in [16]. Nevertheless the classical medial axis, although popular in pattern recognition applications (see e.g. [17,18,19]), is extremely sensitive to oscillations in the boundary of Ω. For example, the effect of small bumps in the contour may be readily evinced from Figure 3. To achieve stability in the computation of the medial axis, in [9] a decreasing filtration of the medial axis has been presented. More precisely, for any positive number λ, one defines the λ-medial axis $\mathcal{M}_\lambda$ as a subset of $\mathcal{M}$. Roughly speaking, $\mathcal{M}_\lambda$ is obtained considering two points on the boundary of Ω indistinguishable if their distance does not exceed λ. Thus, λ may be considered as the optical resolution of the lens used by the observer, thus giving rise to a multi-scale representation of the medial axis.

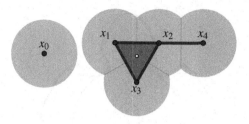

Fig. 2. The dual complex D_α for a collection of 5 points. The balls are depicted in light blue and subdivided into Voronoi regions; the dual complex consists in the 5-vertices, 4 1-simplices and 1 2-simplex depicted in dark blue.

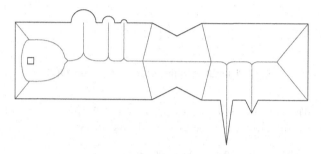

Fig. 3. Example of a 2-dimensional domain and its medial axis

More precisely, for $x \in \mathbb{R}^n$ define $T(x)$ to be the set of points in $\partial\Omega$ at which the distance from x achieves its minimum:

$$T(x) := \{y \in \partial\Omega \,|\, d(x,y) = \min_{z \in \partial\Omega} d(x,z)\}$$

Being Ω bounded, the boundary $\partial\Omega$ is compact; therefore the function $d(x,\cdot)$ achieves its minimum in $\partial\Omega$ at least in one point, i.e. the cardinality $\#T(x)$ is ≥ 1.

The medial axis $\mathcal{M}$ of Ω is precisely the set of points $x \in \Omega$ having at least two closest points:

$$\mathcal{M} := \{x \in \Omega \,|\, \#T(x) \geq 2\}$$

Let $R : \Omega \to \mathbb{R}^+$ be the distance from the boundary function (see Figure 4):

$$R(x) := \min_{z \in \partial\Omega} d(x,z)$$

Notice that the function is strictly positive since Ω is open; actually it is easy to show that R is a short map, i.e. it is 1-Lipschitz.

We may also encode how much the points in $T(x)$ are far apart. A good estimate is the radius of the smallest ball containing them; let thus for $x \in \Omega$:

$$\Lambda(x) = \inf\{r \,|\; \exists c \in \mathbb{R}^n \text{ s.t. } \bar{B}_{c,r} \supseteq T(x)\}$$

We keep also track of the center $c(x)$ of the minimal ball containing $T(x)$. When $x \notin \mathcal{M}$ and therefore, by definition, $T(x) = \{z\}$ reduces to a single point, we have then $\Lambda(x) = 0$

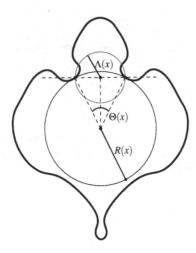

Fig. 4. Pictorial definition of $R(x)$, $\Lambda(x)$, and $\Theta(x)$

and $c(x) = z$. The converse also holds, that is if $x \in \mathcal{M}$ then $\Lambda(x) > 0$. After these notations we may formally define the λ−medial axis $\mathcal{M}_\lambda$ as:

$$\mathcal{M}_\lambda := \{x \in \mathbb{R}^n \mid \Lambda(x) > \lambda\}$$

Clearly $\mathcal{M}_\lambda \supset \mathcal{M}_\mu$ for $\lambda < \mu$ and, in particular, we have $\mathcal{M}_\lambda \subset \mathcal{M}_0 = \mathcal{M}$ for $\lambda > 0$. Roughly speaking $\mathcal{M}_\lambda$ is obtained from the classical medial axis by removing points which have closest points on the boundary not sufficiently far (see Figure 5). In this sense, λ-medial axis may be understood as a cleaning method for the classical medial axis. Nevertheless, persistency may be used again to consider the one-parameter family $(\mathcal{M}_\lambda)_{\lambda \in \mathbb{R}^+}$ at once and thus to analyze the birth and death of topological attributes at various scales.

General Morse Functions. A fundamental source of one-parameter filtrations for applications in shape description and analysis is represented by *Morse functions*, which generalize our previous example about digital terrain elevation. Let X be a n−dimensional manifold (for example a curve or a surface) and let $f : X \in \mathbb{R}$ be a differentiable function. A *critical point* $x \in X$ is a point at which ∇f vanishes; $f(x)$ is then called a *critical value*. A value is called *regular* if it is not critical. A critical point is called *non-degenerate* if the Hessian (i.e. the matrix of second derivatives) is non-singular. Suppose for simplicity that f has a finite number of critical points which are all non-degenerate, and that the critical values are all distinct.

Consider the filtration of X according to the sublevel sets $X_t := f^{-1}(-\infty, t]$. The first result of Morse theory (see e.g. [20]) states that the topology of X_t changes only when t is a critical value of f. Thus, letting $c_1 < c_2 < \ldots < c_k$ be the critical values of f and considering a sequence of regular values $t_0 < \ldots < t_k$ satisfying $t_{i-1} < c_i < t_i$, the topological changes in the family $(X_t)_{t \in \mathbb{R}}$ are summarized by the finite sequence of nested spaces:

$$X_{t_0} \subset X_{t_1} \subset \ldots \subset X_{t_k}$$

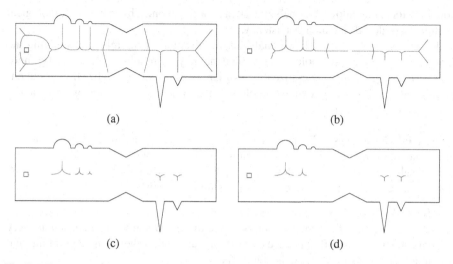

(a)   (b)

(c)   (d)

Fig. 5. Effects of the λ–filtration on the medial axis depicted in Figure 3. From (a) to (d) increasing values of λ are used.

Further, the changes in the topology at a critical point are prescribed by the *index* of the Hessian, i.e. the number of its negative eigenvalues. Namely, if the index of the Hessian at C_i is p, the space X_{t_i} is obtained from $X_{t_{i-1}}$ attaching a disk of dimension p.

Referring again to Figure 1, notice that in our new terminology a minimum correspond to a critical value having index 0 (since the second derivative is positive) while a maximum has index 1. Thus Morse theory tell us that when we meet a minimum, a 0-disk (i.e. a point) should be added, determining the birth of a new connected component.When we meet a maximum, a 1-disk (i.e. an arc) should be attached; the arc is inserted so as to join two connected components. For real digital terrain analysis in dimension 2, the height function may be used as a Morse function; this time, saddles are identified as critical points having index 1 and maxima as critical points with index 2. Of course, to become fully applicable, these results in the "continuous" setting should be translated in the discrete case. Luckily, the results can be transferred to simplicial complexes, that show up in computer and vision as a particular kind of meshes. The category of differentiable functions is then substituted with that of piecewise linear functions. A function of such kind is specified by the values assumed at the vertices of the mesh. Critical points, their degeneracy and indices are formalized through the concepts of lower and upper stars [21]. Moreover, in the discrete setting it is quite easy to extend Morse theory to a more general class of functions. In this way, many interesting functions may be used for shape description purposes via Morse theory; for example the distance from the centroid function may be employed to analyze cavities and protuberances of 3D meshes. See e.g. [22] for further examples.

2.3 Algebraic Topology Tools

Up to now we have illustrated some examples that give rise to filtrations of objects related to computer vision and graphics. In this section, we introduce some algebraic topology tools that allows for analyzing such a family at once, by extracting topological features and identifying their persistency.

Recall that in Section 2.1, we analyzed the filtration of the relief by studying the behavior of the simplest topological invariant, namely the number of connected components. *Homology theory* generalize this invariant to cover and formalize higher dimensional features. We briefly define homology for simplicial complex below, referring to [23] for an introduction.

Simplicial Complexes. Given a set $V = \{v_i\}_{i=1,2,...,p+1}$ consisting in $p+1$ linearly independent points of $\mathbb{R}^n$, the convex hull σ_V of V is called a *p-simplex*. For every proper subset $T \subset V$, the simplex σ_T is a *face* of σ_V. We then say that $\sigma_T < \sigma_V$. A *simplicial complex K* is a finite collection of simplices, satisfying:

– If $\sigma \in K$, then all its faces belong to K
– If $\sigma, \tau \in K$, then their intersection is a face of both σ and τ; this means that every pair of simplices in K can share at most one face and, if they share a point internal to a face, they should share the whole face.

Simplicial Homology. The vectorial space of p–dimensional chains $C_p(K)$ of K over $\mathbb{Z}_2$ is defined as the space of linear combinations of p-simplices with coefficients in $\mathbb{Z}_2 = \mathbb{Z}/2\mathbb{Z}$. The boundary $\partial(\sigma)$ of a p–simplex σ is defined as the sum of its proper faces of maximal dimension:

$$\partial(\sigma) := \sum_{\tau < \sigma;\, \dim \tau = p-1} \tau \in C_{p-1}(K)$$

Extending by linearity, we have a linear map $\partial : C_p(K) \to C_{p-1}(K)$. Its kernel $Z_p(K)$ is a subspace of $C_p(K)$ known as the $p-cycles$. Roughly speaking, it consists in the chains whose simplices join together to enclose a cavity. For example, consider the simplicial complex with three 1-simplices given by $(v_1,v_2),(v_2,v_3),(v_1,v_3)$ and consider the 1-chain $c = (v_1,v_2) + (v_2,v_3) + (v_1,v_3)$. Geometrically c represents the boundary of a triangle. Under the boundary map $\partial c = 0$, since every vertex appears twice in the expansion.

The image $B_{p-1}(K) = \partial(C_p K)$ is called the space of $(p+1)$-*boundaries*. It is easy to see that $\partial^2 = \partial \circ \partial = 0$. For example, let $c' = (v_1,v_2,v_3)$ be the 2-simplex (a triangle) having vertices v_1, v_2, v_3. Then:

$$\partial^2(c') = \partial(c) = 0$$

We may say in this case that the 2-chain c' "fills" the hole individuated by the 1-cycle c. In general we have thus $B_p(K) \subset Z_p(K)$. The p-th homology group $H_p(K)$ measures the p-cycles that are not already filled by a $(p+1)$-chain. More formally, it is defined as the *quotient vector space*:

$$H_p(K) = Z_p(K)/B_p(K)$$

It is now easy to see the relation of H_0 with the number of connected components. Notice that $Z_0(K) = C_0(K)$ since the boundary map is null in dimension zero (for there are no faces in dimension -1). So, we have to study when two vertices v, w are equivalent up to $B_0(K)$. Notice that if there is a sequence of 1-simplices $\sigma_1, \ldots \sigma_s$ connecting v to w than $v = w + \partial(\sum_{i=1}^{s} \sigma_i)$ and thus they define the same class in $H_0(K)$. Viceversa if v and w are not connected by a sequence of 1-simplices, no relation links them inside $H_0(K)$. Thus, a set of generators for $H_0(K)$ is obtained picking up one vertex for each connected component of K. The group $H_1(K)$ is related to the *tunnels* inside K, while $H_2(K)$ is related to the *voids* enclosed by simplices in K; higher homology groups may be similarly interpreted as higher dimensional holes.

Induced Maps. Let be given two simplicial complexes K, K' and a simplicial map $f : K \rightarrow K'$, i.e. a map sending simplices in K to simplices in K' and preserving the face relations. We have an induced map $f_\# : C_p(K) \rightarrow C_p(K')$ at the chain level, which is simply obtained by extending by linearity the map $K \ni \sigma \mapsto f(\sigma) \in K'$. It is easy to show that $\partial \circ f_\# = f_\# \circ \partial$. Thus we have $f_\# Z_p(K) \subset Z_p(K')$ and $f_\# B_p(K) \subset B_p(K')$, so that we have an induced map of vector spaces at the homology level $f_* : H_p(K) \rightarrow H_p(K')$. Notice that by definition the kernel of f_* are the cycles in K that become trivial in K', while the cokernel $H_p(K')/f_* H_p(K)$ is generated by the cycles in K' that are not reached by the cycles in K via f_*. Let $K \subset K'$ and $i : K \hookrightarrow K'$ be the inclusion map. Then, $\ker i_*$ corresponds to the topological features of K that are annihilated when considered in the bigger space K'. Viceversa, $\operatorname{coker} i_*$ corresponds to new topological features which do not show up inside K.

Homology and Filtrations. In Section 2.2 we have shown that, in many practical situations, an object X related to computer vision and graphics may be conveniently approximated via a sequence of nested spaces $(X_t)_{t \in \mathbb{R}}$. Our issue now is to show how such kind of decomposition may lead to a better understanding of the genesis and lifetime of topological features. Recall that topological changes in $(X_t)_{t \in \mathbb{R}}$ occur only at finitely many values of $t = c_1, \ldots, c_k$ with $c_1 < \ldots < c_k$; choosing $t_0 < \ldots < t_k$ such that $t_{i-1} < c_i < t_i$, topological changes in the family are summarized by the sequence of inclusions:

$$X_{t_0} \xrightarrow{i_0} X_{t_1} \xrightarrow{i_1} \ldots \xrightarrow{i_{k-2}} X_{t_{k-1}} \xrightarrow{i_{k-1}} X_{t_k} \tag{1}$$

Using the concept of induced maps, we may pass to the homology level in dimension p $(p = 0, 1, \ldots)$:

$$H_p(X_{t_0}) \xrightarrow{i_{0,*}} H_p(X_{t_1}) \xrightarrow{i_{1,*}} \ldots \xrightarrow{i_{k-2,*}} H_p(X_{t_{k-1}}) \xrightarrow{i_{k-1,*}} H_p(X_{t_k}) \tag{2}$$

This sequence of linear maps encodes the genesis and lifetime of topological features, basically by the analysis of kernel and cokernel of the various maps. For example, let $z \in H_p(X_{t_1})$ whose class in $\operatorname{coker} i_{0,*}$ is not zero. Then, we may say that z is created going from X_{t_0} to X_{t_1}, since it does not exists in X_{t_0}. Then, we may track z along the filtrations, by considering the sequence of its images $i_{0,*}(z), i_{1,*} \circ i_{0,*}(z), \ldots$

If at some point in the sequence the image of z is zero, this means that the topological feature individuated by z dies there. This is good, but it is not sufficient for our purposes. Indeed recall that in the basic example described in Section 2.1 we had to treat the merging of two connected components. Actually if z_1, z_2 are representative cycles for two connected components in X_{t_s} which are merged to a single connected component in $X_{t_{s+1}}$ having w as representative cycle, we have clearly $i_{s,*}(z_1) = i_{s,*}(z_2) = w$. Thus no one of the topological features is annihilated, though they are identified at time $s+1$, since indeed $i_{s,*}(z_1 - z_2) = 0$. Notice that if we choose $(z_1, z_2 - z_1)$ instead of (z_1, z_2) as generators of $H_0(X_{t_s})$, we would discover that one generator dies at time $s+1$ while the other is left untouched. This discussion shows that it is important to find out a suitable global sets of generators for homology groups, in order to analyze persistency in a meaningful manner. Actually, in the basic example (Section 1) we were able to define a simple rule for choosing which connected components should be annihilated. Such rule could be easily restated as a way of choosing a suitable basis for the zeroth homology group. However, in more complex situations and in higher dimensions, it is less clear how to select a suitable basis for homology groups. Nevertheless, rather surprisingly, an algebraic classification result provides a clear answer to this problem [24,25]. We introduce first some notations. Consider the auxiliary vector space obtained by performing the *direct sum* of the homology group shown in Equation 2:

$$M := \bigoplus_{s=0}^{k} H_p(X_{t_s}) \oplus \bigoplus_{s=k+1}^{+\infty} H_p(X_{t_k}) \tag{3}$$

M may be considered as a *graded* vector space, in which for $0 \leq s \leq k$ the part $M^{(s)}$ in degree s is given by $H_p(X_{t_s})$, while for $s > k$ it is given by a copy of $H_p(X_{t_k})$. Thus M ends with an infinite tail corresponding to the stationarity of the filtration after the k-th step. The induced maps $(i_{s,*})_{s=0,1,\dots}$ in homology may be summarized in a map $T : M \to M$ of graded vector spaces that increases the degree by 1. Namely for $m \in M^{(s)}$, we define $T(m) := i_{s,*}(m) \in M^{(s+1)}$, where it is understood that $i_{s,*}$ is the identity map for $s \geq k$. Since T maps M to itself, iterations $(T^2, T^3, \dots)$ and linear combinations thereof do make sense. In particular, any polynomial $p(T) \in \mathbb{Z}_2[T]$ defines a map $M \to M$ or, in other words, we have defined a $\mathbb{Z}_2[T]$-module structure on M. From a mathematical point of view $\mathbb{Z}_2[T]$ is a principal ideal domain; further, in non-pathological situation, M is finitely generated as a $\mathbb{Z}_2[T]$-module. Under these hypotheses, a theorem of algebra (actually a sort of generalization of the existence of the Jordan form for matrices [26]) guarantees that:

$$M \cong \mathbb{Z}_2[T]^\alpha \oplus \bigoplus_{j=1,\dots,r} \left(\frac{\Sigma^{\gamma_j} \mathbb{Z}_2[T]}{T^{\beta_j}} \right) \tag{4}$$

where Σ^{γ_j} denotes degree shifting by the integer γ_j. In other words, the first summand on the right hand side of Equation 4 says that we have α homology classes having infinite life; indeed if m is one generator of $\mathbb{Z}_2[T]^\alpha$, we have that $m, Tm, T^2m, \dots$ are all non zero. The second summand describes homology classes with finite life-time. For example the summand $((\Sigma^{\gamma_j} \mathbb{Z}_2[T])/T^{\beta_j})$ has no components in degree $< \gamma_j$. This means that it describes a class which is born at time t_{γ_j}. Further the summand has components until degree $\beta_j - 1$, i.e. the class is alive at time t_{β_j-1} and dies at time t_{β_j}. Notice that Equation 4 says that we can choose a consistent bases for homology groups across the whole filtrations is such a way to meaningful describe persistency of topological attributes.

The decomposition in Equation 4 may be used to represent pictorially persistency via barcode diagrams; see e.g. [3].

3 Multidimensional Persistency

Up to now we have considered persistency in 1-parameters families of spaces, where a complete answer was given by Equation 4. It is tempting to extend these ideas to more general situations, where the object to be analyzed may be parameterized along multiple geometric dimensions.

Multiple Morse Functions. A general source of such kind of *multi-filtrations* is represented by manifolds equipped with several Morse functions. For example, let X be a manifold and let f_1, $f_2 : X \to \mathbb{R}$ be two Morse functions. For $(t_1, t_2) \in \mathbb{R}^2$, we may consider the space:

$$M_{(t_1,t_2)} := f_1^{-1}(-\infty, t_1] \cap f_2^{-1}(-\infty, t_2] \tag{5}$$

i.e. the intersection of the sublevel sets relative to f_1 and f_2. Consider the following partial order on $\mathbb{R}^2$: for (t_1, t_2), $(t_1', t_2') \in \mathbb{R}^2$, we define $(t_1, t_2) \preceq (t_1', t_2')$ iff $t_1 \leq t_1'$ and $t_2 \leq t_2'$. Notice that for $(t_1, t_2) \preceq (t_1', t_2')$ we have an inclusion $X_{(t_1,t_2)} \hookrightarrow X_{(t_1',t_2')}$. Thus the multi-filtration is organized over $\mathbb{R}^2$ with the order $\preceq$.

The (λ, θ)-Medial Axis. In Section 2.1 we have defined the λ-filtration of the medial axis which is useful to remove noise and achieve stability in its computation. We now define a second filtration which can capture geometrical features instead, like sharp corner or the presence of parallel segments in the boundary of a domain Ω. The filtrations behave thus in different manner and both are therefore of interest. We may attempt to fuse the information extracted from both filtrations in the framework of multidimensional persistence. To define the θ−filtration, let's consider, with the previous notation, the function Θ defined for $x \in \Omega$:

$$\Theta(x) = \begin{cases} 2\arcsin\left(\frac{\Lambda(x)}{R(x)}\right) & \text{if } \#T(x) = 2 \\ 0 & \text{otherwise} \end{cases}$$

where arcsin takes values into $(-\pi/2, \pi/2]$. Note that, when $\#T(x) = 2$, the value $\Theta(x)$ is the angle by which the points in $T(x)$ are seen from the point x (see Figure 4 for a pictorial explanation of the function Θ). We may now filter the medial axis $\mathcal{M}$ with the upper level sets of the function Θ. The effect of the θ−filtration may be appreciated in Figure 6. The $\lambda-$ and $\theta-$ filtrations may be now coupled as in the case of several Morse functions above.

3.1 Algebraic Topology Tools

In [10] an attempt to the study of multidimensional persistence has been presented; the authors use multi-parameter filtrations organized over a grid of points in $\mathbb{R}^n$ with a suitable partial order. Then, they are able to define an action of the polynomial ring $\mathbb{Z}_2[T_1, \ldots, T_n]$ in several variables on a module M that generalizes the module defined

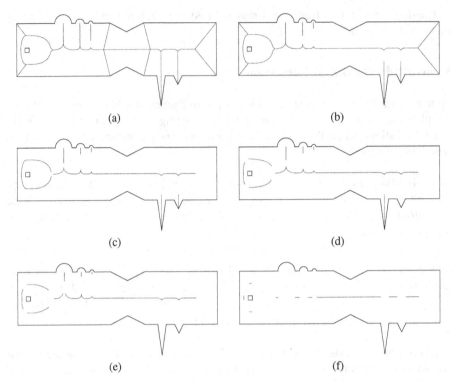

Fig. 6. Effects of the θ−filtration on the medial axis depicted in Figure 3. From (a) to (f) increasing values of θ are used.

in Equation 3. However, since $\mathbb{Z}_2[T_1, \ldots, T_n]$ is not a principal ideal domain, the decomposition result into elementary pieces (Equation 4) is no more available. Actually, their analysis identifies some continuous invariant for multidimensional persistency, thus showing that general persistency cannot be described solely in terms of discrete invariants (i.e. birth/death times). In [27], the authors address the same kind of multi-filtrations and present a reduction to the one-dimensional case, by defining a suitable *foliation* of $\mathbb{R}^n$.

In the subsections below, we extend this analysis to more general situations, in which the multi-parameters filtration is organized over an arbitrary poset. The module M in Equation 3 is replaced by a *sheaf* over the poset. Also in this general setting, we are able to define discrete invariants corresponding to the instant of creation and lifetime.

Posets. Let $(P, \preceq)$ be a *partially ordered set* –a *poset* for short– i.e. a set P endowed with a binary relation $\preceq$:

$$x \preceq y$$

which is reflexive, antisymmetric and transitive.

For example, consider the space $\mathbb{R}^n$. For $x = (x_1, \ldots, x_n)$ and $y = (y_1, \ldots, y_n) \in \mathbb{R}^n$ we define:

$$x \preceq y \quad \Leftrightarrow \quad x_i \leq y_i \quad \forall i = 1, \ldots, n$$

Clearly $\preceq$ is reflexive, antisymmetric and transitive and, thus, $(\mathbb{R}^n, \preceq)$ is a poset.

Notice that any subset T of a poset P inherits a structure of poset. For $x, y \in P$, we say that is *covered* by y if for any $z \in P$, $z \neq y$ such that $x \preceq z \preceq y$ we have $z = x$; we then write $x \lhd y$. An *ideal* I of P is a subset of P such that $x \in I$, $y \in P$, $y \preceq x \Rightarrow y \in I$, i.e. if I contains an element x, it also contains all the elements less than x. For $x \in P$, the set $I_x = \{y \in P \mid y \preceq x\}$ is the ideal *generated* by x. Notice that I_x has x as unique maximum element. A *filter* is the dual notion of an ideal; in particular we define $I^x = \{y \in P \mid y \succeq x\}$ as the filter generated by x.

Recall that a *category* is a collection of objects and arrows (or morphisms) connecting them. A poset $(P, \preceq)$ is equivalent to the small category C_P having as objects

$$Ob(C_P) = P$$

i.e. the points in the poset, and having as arrows connecting them:

$$
\begin{aligned}
\mathrm{Hom}(x, y) &= \{(x, y)\} && \text{if } x \preceq y \\
&= \emptyset && \text{otherwise}
\end{aligned}
$$

i.e. two object $x, y \in P$ are connected by an arrow iff $x \preceq y$; in that case, the arrow points from x to y and it is unique.

Diagrams of Spaces. We may use a poset to organize both a collection of topological spaces and the maps among them, by using the notion of *diagram of spaces*. A diagram of spaces $\mathcal{D}$ over $(P, \preceq)$ is a collection of topological spaces together with a collection of continuous maps:

$$\{A_x \mid x \in P\}; \qquad\qquad \{f_{x,y} : A_x \to A_y \mid x \preceq y\}$$

satisfying:

$$f_{x,x} = id_{A_x}; \qquad\qquad x \preceq y \preceq z \Rightarrow f_{x,z} = f_{y,z} \circ f_{x,y}.$$

In particular, the above conditions say that the maps in the diagram fit together according to the order in the poset. Formally, a diagram of spaces is a *functor* from C_P to the category of topological spaces.

We will only consider the case when all the maps $f_{x,y}$ are inclusion maps of topological spaces. Let Q_k be the poset of the integers $i : 0 \leq i \leq k$ with the natural order. Then a diagram of spaces over Q_k is equivalent to a one-parameter filtration of a topological space (see Eq. 1). If Q_{k_1,k_2} is the poset of pairs of integers placed on a $k_1 \times k_2$-grid we obtain a multi-filtration suitable for the analysis of multiple Morse functions (see Eq. 5) or the (λ, θ)−medial axis.

Sheaves Over Posets. We may formalize the interdependencies among the topological features of the spaces in a diagram of spaces by the concept of *sheaf* over a poset [28,29]. A sheaf $\mathcal{F}$ over $(P, \preceq)$ is a collection of vector spaces $\{\mathcal{F}_x\}$ together with a collection of linear maps among them:

$$\{\mathcal{F}_x \mid x \in P\}; \qquad\qquad \{T_{x,y} : \mathcal{F}_x \to \mathcal{F}_y \mid x \preceq y\}$$

satisfying:

$$T_{x,x} = id_{\mathcal{F}_x}; \qquad\qquad x \preceq y \preceq z \Rightarrow T_{x,z} = T_{y,z} \circ T_{x,y}.$$

Formally, a sheaf is a *functor* from C_P to the category of vector spaces.

Sheaves over posets arise in the analysis of persistency from diagram of spaces through the use of homology theory. Actually given a diagram of spaces $\mathcal{D}$ passing to the homology level in dimension p, we get the sheaf $\mathcal{H}_p(\mathcal{D})$:

$$\{H_p(A_x) \,|\, x \in P\}; \qquad\qquad \{f_{x,y;*} : H_p(A_x) \to H_p(A_y) \,|\, x \preceq y\}$$

This construction generalizes the sequence in Eq. 2, which is obtained by considering the poset Q_k.

3.2   Discrete Invariants

Let $\mathcal{D}$ be a diagram of spaces over P and $\mathcal{F} = \mathcal{H}_p(\mathcal{D})$ be the sheaf obtained by considering the homology in dimension p.

Instant of Creation. For an element $x \in P$, we may study which topological features are born at A_x; let I_x be the ideal generated by x. The sheaf restricted to the subposet I_x contains all the "past" of x for what regards topological features. Actually, no other element outside I_x has an arrow with target x. For $y \in P$, the topological features of A_y -seen inside A_x- are given by the image of the homology map $f_{y,x;*}(H_p(A_y))$. Thus new topological features are described by the cokernel:

$$\bar{H}_p(A_x) = H_p(A_x) / (\underset{y \in I_x}{+} f_{y,x;*}(H_p(A_y)))$$

Since all the involved maps factor through elements covered by x, it is easy to show that:

$$\bar{H}_p(A_x) = H_p(A_x) / (\underset{y \lhd x}{+} f_{y,x;*}(H_p(A_y))),$$

thus simplifying the computation of $\bar{H}_p(A_x)$. A discrete invariant for the new topological features is represented by the rank, i.e. we may define

$$\mu_x = \mathrm{rank}(\bar{H}_p(A_x))$$

as the extent of new topological features created at time x.

Lifetime. The lifetime of a topological feature may be tracked in a similar manner. Notice that given $x \in P$ the "future" of x is represented by the sheaf restricted to the filter I^x generated by x. Further, we may factor out the past of x, considering a new sheaf $\mathcal{F}(x)$ on I^x. For $y \in I^x$, the vector space $\mathcal{F}(x)_y$ is defined as:

$$\mathcal{F}(x)_y = H_p(A_y) / (\underset{z \in I_x}{+} f_{z,y;*}(H_p(A_z)))$$

The maps connecting the vector spaces in the sheaf $\mathcal{F}(x)$ are, by definition, the induced ones.

The persistency of features born at time x until time $y \in I^x$ is then described by the rank $\rho_{x,y}$ of the following linear map:

$$\bar{f}_{x,y;*} : \mathcal{F}(x)_x \to \mathcal{F}(x)_y$$

Notice that the creation invariant μ_x is a special case of $\rho_{x,y}$; indeed $\mu_x = \rho_{x,x}$. Further, notice that these definitions are compatible with the heuristic rule described in Section 2.1.

4 Conclusions

In this paper, after motivating the use of persistency for robust shape description, we introduced the basic results of persistency theory for one-parameter filtrations and, then, we described algebraic topology tools for the analysis of more general filtrations. Further work will focus on methods for the hierarchical organization and exploitation of such features.

References

1. Edelsbrunner, H., Harer, J.: Persistent homology — a survey. In: Twenty Years After (AMS) (to appear)
2. Collins, A., Zomorodian, A., Carlsson, G., Guibas, L.: A barcode shape descriptor for curve point cloud data. Computers and Graphics 28, 881–894 (2004)
3. Carlsson, G., Zomorodian, A., Collins, A., Guibas, L.J.: Persistence barcodes for shapes. International Journal of Shape Modeling 11, 149–187 (2005)
4. Biasotti, S., Giorgi, D., Spagnuolo, M., Falcidieno, B.: Size functions for 3D shape retrieval. In: SGP 2006: Proceedings of the fourth Eurographics Symposium on Geometry Processing, Aire-la-Ville, Switzerland. Eurographics Association, Switzerland, pp. 239–242 (2006)
5. Biasotti, S., Giorgi, D., Spagnuolo, M., Falcidieno, B.: Size functions for comparing 3D models. Pattern Recognition (2008), doi:10.1016/j.patcog.2008.02.003
6. Frosini, P., Landi, C.: Size theory as a topological tool for computer vision. Pattern Recognition and Image Analysis 9, 596–603 (1999)
7. Li, X., Hu, C., Liang, J.: Simplicial edge representation of protein structures and alpha contact potential with confidence measure. Proteins 53, 792–805 (2003)
8. Zomorodian, A., Guibas, L.J., Koehl, P.: Geometric filtering of pairwise atomic interactions applied to the design of efficient statistical potentials. Computer Aided Geometric Design 23, 531–544 (2006)
9. Chazal, F., Lieutier, A.: The "lambda-medial axis". Graphical Models 67, 304–331 (2005)
10. Carlsson, G., Zomorodian, A.: The theory of multidimensional persistence. In: Proc. ACM Symposium of Computational Geometry, pp. 184–193 (2007)
11. Bernardini, F., Mittleman, J., Rushmeier, H.E., Silva, C.T., Taubin, G.: The ball-pivoting algorithm for surface reconstruction. IEEE Trans. Vis. Comput. Graph. 5, 349–359 (1999)
12. Edelsbrunner, H., Mücke, E.P.: Three-dimensional alpha shapes. ACM Trans. Graph. 13, 43–72 (1994)
13. Zomorodian, A.: Computing and Comprehending Topology: Persistence and Hierarchical Morse Complexes. PhD thesis, University of Illinois at Urbana-Champaign (2001)
14. Zomorodian, A.: Topology for Computing. Cambridge University Press, New York (2005)

15. Delaunay, B.: Sur la sphere vide. Otdelenie Matematicheskii i Estestvennyka Nauk 7, 793–800 (1934)
16. Lieutier, A.: Any open bounded subset of $\mathbb{R}^n$ has the same homotopy type as its medial axis. Computer-Aided Design 36, 1029–1046 (2004)
17. Matheron, G.: Examples of topological properties of skeletons & On the negligibility of the skeleton and the absolute continuity of erosions. In: Image Analysis and Math. Morphology: Theoretical Advances, vol. 2, pp. 216–256. Academic Press, London (1988)
18. Blum, H.: A transformation for extracting new descriptions of shape. In: Computer Methods in Images Analysis, pp. 153–171 (1977)
19. Blum, H., Nagel, R.: Shape description using weighted symmetric axis features. Pattern Recognition 10, 167–180 (1978)
20. Milnor, J.: Morse theory. Princeton University Press, Princeton (1963)
21. Banchoff, T.F.: Critical points and curvature for embedded polyhedral surfaces. Am. Math. Monthly 77, 465–485 (1970)
22. Attene, M., Biasotti, S., Mortara, M., Patanè, G., Spagnuolo, M., Falcidieno, B.: Topological, geometric and structural approaches to enhance shape information. In: Eurographics Italian Chapter Conference, pp. 7–13 (2006)
23. Hatcher, A.: Algebraic Topology. Cambridge University Press, Cambridge (2001)
24. Zomorodian, A., Carlsson, G.: Computing persistent homology. In: Proc. ACM Symposium on Computational Geometry, pp. 347–356 (2004)
25. Zomorodian, A., Carlsson, G.: Computing persistent homology. Discrete & Computational Geometry 33, 249–274 (2005)
26. Lang, S.: Undergraduate Algebra. Springer, Heidelberg (1990)
27. Cerri, A., Biasotti, S., Giorgi, D.: k-dimensional size functions for shape description and comparison. In: ICIAP, pp. 795–800 (2007)
28. Baclawski, K.: Whitney numbers of geometric lattices. Advances in Mathematics 16, 125–138 (1975)
29. Yuzvinsky, S.: Cohen-Macaulay rings of sections. Advances in Mathematics 63, 172–195 (1987)

Novel Computerized Methods in System Biology – Flexible High-Content Image Analysis and Interpretation System for Cell Images

Petra Perner

Institute of Computer Vision and applied Computer Sciences, IBaI
Arno-Nitzsche-Str. 43, 04277 Leipzig
pperner@ibai-institut.de
www.ibai-institut.de

Abstract. In the rapidly expanding fields of cellular and molecular biology, fluorescence illumination and observation is becoming one of the techniques of choice to study the localization and dynamics of proteins, organelles, and other cellular compartments, as well as a tracer of intracellular protein trafficking. The automatic analysis of these images and signals in medicine, biotechnology, and chemistry is a challenging and demanding field. Signal-producing procedures by microscopes, spectrometers and other sensors have found their way into wide fields of medicine, biotechnology, industrial and environmental analysis. With this arises the problem of the automatic mass analysis of signal information. Signal-interpreting systems which automatically generate the desired target statements from the signals are therefore of compelling necessity. The continuation of mass analysis on the basis of the classical procedures leads to investments of proportions that are not feasible. New procedures and system architectures are therefore required. We will present, based on our flexible image analysis and interpretation system *Cell_Interpret*, new intelligent and automatic image analysis and interpretation procedures. We will demonstrate it in the application of the HEp-2 cell pattern analysis.

Keywords: Image Analysis and Interpretation, High-Content Analysis of Images HCA, Automation and Standardization of Visual Inspection Tasks, Image-Mining, Systems for Knowledge Discovery and Interpretation, Microscopic Cell Image Analysis.

1 Introduction

In the rapidly expanding fields of cellular and molecular biology, fluorescence illumination and observation is becoming one of the techniques of choice to study the localization and dynamics of proteins, organelles, and other cellular compartments, as well as a tracer of intracellular protein trafficking.

Quantitative imaging of fluorescent proteins and patterns is accomplished with a variety of techniques, including wide-field, confocal and multiphoton microscopy, ultrafast low-light level digital cameras and multitracking laser control systems. These microscopic images can be of 2-dimensional or 3-dimensional nature, or even videos recording the life cycle of a cell.

P. Perner and O. Salvetti (Eds.): MDA 2008, LNAI 5108, pp. 139–157, 2008.
© Springer-Verlag Berlin Heidelberg 2008

Currently the interpretation of the resulting pattern in these digital images is usually done manually. However the huge amount of data created and the growing use of these techniques in industry for pharmacological aspects or diagnostic purposes in medicine require automatic image interpretation procedures. These image interpretation procedures should allow to interpret these images automatically, and also to detect automatically new knowledge to study the cellular and molecular processes.

The continuation of mass image analyses on the basis of the classical procedures leads to investments of proportions that are not feasible. New procedures based on image mining and case-based reasoning are therefore required.

We are developing methods that allow the automatic analysis of these images for the discovery of patterns, new knowledge and relations. The present work is applied to 2-dimensional microscopic fluorescent images, but will be continued with 3-d-image and video analysis. The aim of our work is to provide the system with image-analysis, feature-extraction and knowledge-discovery functions that are suited for mining a set of microscopic cell images for the automatic detection of image-interpretation knowledge and then applying this knowledge within the same system for automatic image interpretation of the HEp-2 cell images. At the end the system can work on-line in a pharmaceutical drug discovery process or in a medical laboratory process and automatically interpret the patterns on the cells in the image and calculate quantitative information about the cell pattern.

The developed processing functions should make the system flexible enough to deal with different kinds of cell-images and different image qualities and require a minimal number of interactions with the user for knowledge mining. The image-interpretation process is running fully automatically, based on the image-analysis and feature-extraction procedures developed for this kind of image analysis and the learned interpretation knowledge by the developed knowledge-mining procedures.

2   Challenges and Requirements to the Systems

Application-oriented systems that can only solve one specific task are very costly and it takes time to develop them. The success of automatic image-interpretation systems can only be guaranteed when the development effort is as low as possible and when they can be adapted quickly to different needs and tasks.

It is preferable that the automatic system not only calculates image features from the images but also maps the measurements to the desired information the user wants to obtain with his experiment. This views High-Content Image Analysis as a pattern recognition and image interpretation problem rather than as an image measurement problem where all possible image features are extracted from the images for further analysis. The pattern or the final information, such as e.g. "do the bacteria co-localize with the lysosomes", is the central focus of the image analysis and the system should provide all functions that are necessary to achieve this result.

That requires developing systems that can run on a class of applications such as microscopic fluorescent images. Such systems should have functions that are able to:

- automatically detect single cells in the image regardless of the image quality with high accuracy, robustness, the ability for reproduction, and flexibility,

- automatically describe the properties of the cell nucleus and the cytoplasm by image features (numerical and symbolical),
- automatically interpret the images into cell patterns or other decisions (prediction),
- automatically detect new knowledge from image data and apply it to automatic interpretation.

The challenges are:

- New strategies are necessary that are able to adapt the system to changing environmental conditions during image capture, user needs and process requirements.
- Introduction of Case-Based-Reasoning (CBR) strategies and Data-Mining strategies [1] into image-interpretation systems on both the low-level and high-level to satisfy these requirements.

3 The Architecture

Our answer to this problem is a system architecture [2] named *Cell_Interpret* (see Figure 1) that is comprised of two main parts:

- the on-line part that is comprised of the image analysis and the image interpretation part.
- the off-line part that is comprised of the database and the data mining and knowledge discovery part.

These two units communicate over a database of image descriptions, which is created in the frame of the image-processing unit. This database is the basis for the image-mining unit.

The on-line part can automatically detect objects, extract image features from the objects and classify the recognized objects into the respective classes based on the prior stored decision rules. The interface between the off-line and the on-line part is the database where images and calculated image features are stored. The off-line part can mine the images for a prediction model or discover new groups of objects, features or relations. These similar groups can be used for learning the classification model or just for understanding the domain. In the later case the discovered information is displayed on the terminal of the system to the user. Once a new prediction model has been learnt the rules are inputted into the image interpretation part for further automatic interpretation after approval of the user.

Besides that there is an archiving and management part that controls the whole system and stores information for long-term archiving.

Images can be processed automatically or semi-automatically. In the first case, a set of images specified by the expert is automatically segmented into background and objects of interest and the feature extraction procedures installed in the image analysis system are used for each object to automatically calculate all features. All features are extracted regardless of their applicability for the specific application. This requires executing feature subset selection methods later on. For semi-automatic processing, an image from the image archive is selected by the expert and then is it displayed on the monitor. To perform image processing an expert communicates with a computer. In this mode he has the option to calculate features based on the feature extraction procedures and/or record symbolic features based on his expert knowledge. This procedure ensures that also complicated image features, which are difficult to name, articulate or develop automatic feature extraction procedures, can also be taken into account and further evaluated by image mining. After the feature has been established by evaluating the acquired data base, the proper automatic feature extraction procedure can be developed and included into the system and made available for High-Content Analysis. The intelligence of the system will therefore incrementally improve.

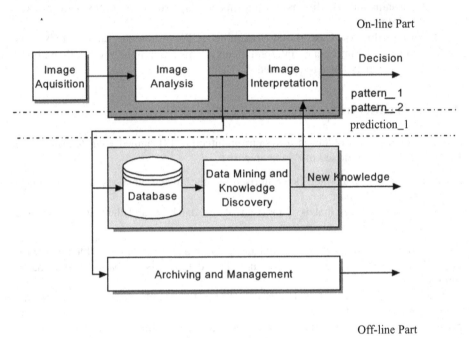

Fig. 1. Architecture of *Cell_Interpret*

4   Case-Based Image Segmentation

Image segmentation is a process of dividing an image into a number of different regions such that each region is homogeneous with respect to a given property, but the union of any two adjacent regions is not.

Image thresholding is a well-known technique for image segmentation. Because of its wide applicability to many areas of digital image processing, a large number of thresholding methods have been proposed over the years (see, e.g., [3-5]). Image thresholding has low computational complexity, which makes it an attractive method, but does not take into account spatial information and is mostly suitable for images where the gray-levels constitute well defined peaks, separated by not too broad and flat valleys.

Another common approach to image segmentation is based on feature space clustering, which has sometimes been regarded as the multidimensional extension of the concept of thresholding. Clustering schemes using different kinds of features (multi-spectral information, mean/variation of gray-level, texture, color) have been suggested (see, e.g., [6-8]). This approach can be successfully used if each perceived region of the image constitutes an individual cluster in the feature space. This requires a careful selection of the proper features, which depends on image domain.

Segmentation can also be accomplished by using region-based methods, or edge-detection-based methods, or methods based on a combination of those two approaches (see, e.g., [9-11]). Region-based methods imply the selection of suitable seeds from which to perform a growing process. In general, region merging and region splitting are accomplished to obtain a meaningful number of homogeneous regions. Seed selection and homogeneity criterion play a critical role for the quality of the obtained results. Edge-detection-based methods follow the way in which human observers perceive objects, as they take into account the difference in contrast between adjacent regions. Edge detection does not work well if the image is not well contrasted, or in the presence of ill-defined or too many edges.

Watershed-based segmentation (see, e.g., [12]) exploits both region-based and edge-detection-based methods. The basic idea of watershed-based segmentation is to identify in the gray-level image a suitable set of *seeds* from which to perform a growing process. If the main feature taken into account is gray-level distribution, the seeds are mostly detected as the sets of pixels with locally minimal gray-level (called *regional minima*). The growing process groups each seed with all pixels that are closer to that seed than to any other seed, provided that a certain homogeneity in gray-level is satisfied. Thus, watershed-based segmentation limits the drawbacks of region-based and edge-detection-based methods.

To overcome the drawbacks of the algorithms mentioned above, learning methods are applied to image segmentation. These learning methods are applied to learn the mapping between image features and semantically meaningful parts, to learn the parameters of the segmentation algorithm or to learn the mapping between rank performance of the segmentation algorithm and the image features.

There are statistical learning methods, machine learning methods, neural-net-based learning methods, and learning methods using a combination of different techniques. The main drawbacks of these methods are: 1) the need of a sufficiently large training set, and 2) the need of training again the whole model, when new data come in. Therefore, it seems to be useful to use Case-based Reasoning (CBR) for a flexible image segmentation system, since CBR can be used as a reasoning approach as well as an incremental knowledge-acquisition approach.

We propose a novel image-segmentation scheme based on case-based reasoning. We use CBR for meta-learning of the segmentation parameters (see Section 4.1) and for case-based object recognition (see Section 4.2).

4.1 CBR Meta Learning for Image Segmentation

The case-based reasoning unit for meta learning of image segmentation parameters [13] consists of a case base in which formerly processed cases are stored. A case is comprised of image information, non-image information (e.g. image-acquisition parameters, object characteristics and so on), and image-segmentation parameters. The task is now to find the best segmentation for the current image by looking up the case base for similar cases. Similarity determination is done based on non-image information and image information. The evaluation unit will take the case with the highest similarity score for further processing. In case there are two or more cases with the same similarity score, the case appearing first will be taken. After the closest case has been chosen, the image-segmentation parameters associated with the selected case will be given to the image-segmentation unit and the current image will be segmented (see Figure 2). It is assumed that images having similar image characteristics will show similar good segmentation results when the same segmentation parameters are applied to these images. The image segmentation algorithm is in our case a histogram-based image-segmentation algorithm [13] and a watershed-based image-segmentation algorithm [14].

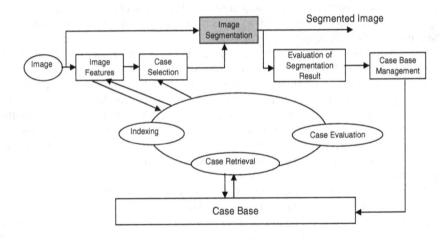

Fig. 2. CBR Image Segmentation Unit

The result of the segmentation process can be observed by the user or an automatic evaluation procedure. When the evaluation is done by the user, he compares the original image with the labeled image on display. If he detects deviations of the marked areas in the segmented image from the object area in the original image, which should be labeled, then he will evaluate the result as incorrect and case-base management will start. This will also be done if no similar case is available in the case-base. The proposed method is close to the critique-modify framework described by Grimnes et al. [15].

The evaluation procedure can also be done automatically. However, the drawback is that there is no general procedure available. It can only be done in a domain-dependent fashion.

Once the chosen evaluation procedure observes a bad result, the respective case is tagged as bad case. The tag describes the critique in more detail.

In an off-line phase, the best segmentation parameters for the image are determined by an optimization procedure and the attributes/features, which are necessary for similarity determination, are calculated from the image. Both, the segmentation parameters and the attributes calculated from the image, are stored into the case-base as a new case. In addition to that the non-image information is extracted from the file header and stored together with the other information in the case-base. During storage, case generalization will be done to ensure that the case base will not become too large.

4.2 Case-Based Object Recognition

We propose our case-based object recognition method to recognize objects by their shape. In contrast to traditional object recognition methods [16] our method is comprised of a case mining part and the object recognition part [17]. The case mining part can learn the desired contour of the object and the number of contours necessary for recognizing a particular class of objects. The learnt contours make up the case base and are the basis for the case-based object-recognition method.

The objects in the image may be occluded, touching, or overlapping. It can also happen that only part of the object appears in the image.

A case-based object-recognition method uses cases that generalize the original objects and matches them against the objects in the image, see Fig. 3. During this procedure a score is calculated that describes the quality of the fit between the object and the case. The case can be an object model which describes the inner appearance of the object as well as its contour. In our case the appearance of the whole object can be very diverse. The shape seems to be the feature that generalizes the objects. Therefore, we decided to use contour models. We do not use the gray values of the model, but instead use the object's edges. For the score of the match between the contour of the object and the case we use a similarity measure based on the scalar product. It measures the average angle between the vectors of the template and the object.

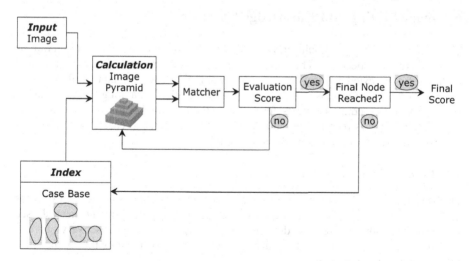

Fig. 3. Principle of case-based object-recognition architecture

The acquisition of the case is done semi-automatically. Prototypical images are shown to an expert. The expert manually traces the contour of the object with the help of the cursor of the computer. Afterwards the number of contour points is reduced for data-reduction purposes by interpolating the marked contour by a first-order polynom. The marked object shapes are then aligned by the Procrustes Algorithm [18]. From the sample points the direction vector is calculated. From a set of shapes general groups of shapes are learnt by conceptual clustering which is a hierarchical incremental clustering method [19]. The prototype of each cluster is calculated by estimating the mean shape [19] of the set of shapes in the cluster and is taken as a case model.

5 Automatic and Symbolic Feature Extraction

The system can now, based on the feature-extraction procedure installed in the system, calculate image features for the labeled objects. These features are composed of statistical gray-level features, the object contour, square, diameter, shape, [20] and a novel texture feature based on random sets [21] that is flexible enough to describe different textures of cells. The system evaluates or calculates image features and stores their values in a database of image features. Each entry in the database presents features of the object of interest. These features can be numerical (calculated on the image) and symbolical (determined by the expert as a result of image reading by the expert). In the latter case the expert evaluates object features according to the attribute list, which has to be specified in advance for object description, or is based on a visual ontology available for visual content description. Then the user feeds these values into the database. When the expert has evaluated a sufficient number of images, the resulting database can be used for the image-mining process.

6 Image Mining and Knowledge Discovery

The image mining part should allow extracting knowledge or making observations from different perspectives. Therefore, we have included methods for predictions and methods for knowledge discovery [1]. Knowledge discovery methods allow us to summarize data into groups and patterns or observe relations among groups. Usually they are prior to prediction. We prefer conceptual clustering [1] for this task since the discovery process is incremental and therefore fits perfectly to case-based reasoning and decision tree induction as prediction methods.

6.1 Decision Tree Induction

Decision tree induction allows one to learn from a set of data samples a set of rules and basic features necessary for decision-making in a specified diagnostic task, see Figure 4. The induction process does not only act as a knowledge discovery process, it also works as a feature selector, discovering a subset of features that is the most relevant to the problem solution.

Decision trees partition decision space recursively into sub-regions based on the sample set.

In this way the decision trees recursively break down the complexity of the decision space. The outcome has a format which naturally presents a cognitive strategy that can be used for the human decision-making process. The rules contained in the tree can be understood by human. Therefore a decision tree is a representation form that has explanation capability.

For any tree all paths lead to a terminal node, corresponding to a decision rule that is a conjunction (AND) of various tests. If there are multiple paths for a given class, then the paths represent disjunctions (ORs).

The developed tool allows choosing different kinds of methods for feature selection, feature discretization, pruning of the decision tree and evaluation of the error rate. It provides an entropy-based measure, a gini-index, gain-ratio and chi square method for feature selection [1].

The following methods for feature discretization are provided: cut-point strategy, chi-merge discretization, minimum description length, principal based discretization method and lvq-based method [1]. These methods allow one to make discretization of the feature values into two and more intervals during the process of decision-tree building. Depending on the chosen method for attribute discretization, the result will be a binary or n-ary tree. The later will lead to more accurate and compact trees.

The tool allows one to chose between cost-complexity pruning, error-reduction-based methods and pruning by confidence-interval prediction. The tool also provides functions for outlier detections.

To evaluate the obtained error rate one can choose test-and-train and n-fold cross validation. Missed values can be handled by different strategies [1].

The user selects the preferred method for each step of the decision tree induction process. After that the induction experiment can start on the acquired database. A resulting decision tree will be displayed to the user. He/she can evaluate the tree by checking the features used in each node of the tree and comparing them with his/her domain knowledge.

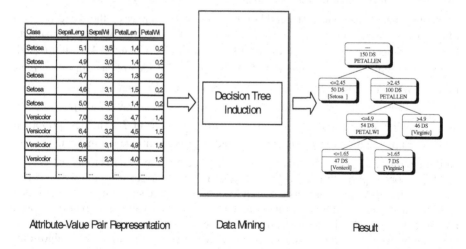

Attribute-Value Pair Representation   Data Mining   Result

Fig. 4. Basic Principle of Decision Tree Induction

Once the diagnosis knowledge has been learnt, the rules are provided either in txt-format or XML format for further use in the image interpretation part or the expert can use the diagnosis component of the tool for interactive work. It has a user-friendly interface and is set up in such a way that non-computer specialists can handle it very easily.

6.2   Case-Based Reasoning for Image Interpretation

It is difficult to apply decision trees in domains where generalized knowledge is lacking. But often there is a need for a prediction system, even though there is not enough generalized knowledge. Such a system should a) solve problems using the already stored knowledge and b) capture new knowledge, making it immediately available to solve the next problem. To accomplish these tasks case-based reasoning is useful. Case-based reasoning explicitly uses past cases from the domain expert´s successful or failing experience.

Therefore, case-based reasoning can be seen as a method for problem-solving as well as a method to capture new experience in an incremental fashion and make it immediately available for problem-solving. It can be seen as a learning and knowledge-discovery approach, since it can capture from new experience some general knowledge such as case classes, prototypes and some higher-level concepts. We find these methods especially applicable for inspection and diagnosis tasks. In the case of these applications people store prototypical images into a digital image catalogue rather than a large set of different images [22].

We have developed a unit for *Cell_Interpret* that can perform similarity determination between cases, as well as prototype selection [23] and feature weighting [24]. We call $x_n \in \{x_1, x_2, ..., x_n\}$ a nearest-neighbor to x if $\min d(x_i, x) = d(x_n´, x)$, where $i = 1, 2, ..., n$. The instance x is classified into category C_n, if x_n is the nearest neighbor to x and x_n belongs to class C_n.

In the case of the k-nearest neighbor we require k-samples of the same class to fulfill the decision rule. As a distance measure we use the Euclidean distance. Prototype Selection from a set of samples is done by Chang`s Algorithm [23]. Suppose a training set T is given as $T = \{t^1, ..., t^m\}$. The idea of the algorithm is as follows: We start with every point in T as a prototype. We then successively merge any two closest prototypes p^1 and p^2 of the same class by a new prototype p, if the merging will not downgrade the classification of its patterns in T. The new prototype p may simply be the average vector of p^1 and p^2. We continue the merging process until the number of incorrect classifications of the patterns in T starts to increase.

The wrapper approach is used for selecting a feature subset from the whole set of features. This approach conducts a search for a good feature subset by using the k-NN classifier itself as an evaluation function. The 1-fold cross-validation method is used for estimating the classification accuracy and the best-first search strategy is used for the search over the state space of possible feature combination. The algorithm terminates if we have not found an improved accuracy over the last k search states. The feature combination that gave the best classification accuracy is the remaining feature subset. After we have found the best feature subset for our problem, we try to further improve our classifier by applying a feature-weighting technique.

The weights of each feature w_i are changed by a constant value δ: $w_i := w_i \pm \delta$. If the new weight causes an improvement of the classification accuracy, the weight will be updated accordingly; if not, the weight will remain as it is. After the last weight has been tested the constant δ will be divided into half and the procedure repeats. The procedure terminates if the difference between the classification accuracy of two iterations is less than a predefined threshold.

6.3 Conceptual Clustering

The intention of clustering as another image mining function is to find groups of similar cases among the data according to the observation. This can be done based on one feature or a feature combination. The resulting groups give an idea how data fit together and how they can be classified into interesting categories.

Classical clustering methods only create clusters but do not explain why a cluster has been established. Conceptual clustering methods build clusters and explain why a set of objects confirm a cluster. Thus, conceptual clustering is a type of learning by observation and it is a way of summarizing data in an understandable manner [1]. In contrast to hierarchical clustering methods, conceptual clustering methods build the classification hierarchy not only based on merging two groups. The algorithmic properties are flexible enough to dynamically fit the hierarchy to the data. This allows incremental incorporation of new instances into the existing hierarchy and updating this hierarchy according to the new instance.

A concept hierarchy is a directed graph in which the root node represents the set of all input instances and the terminal nodes represent individual instances. Internal nodes stand for sets of instances attached to the nodes and represent a super-concept. The super-concept can be represented by a generalized representation of this set of instances such as the prototype, the mediod or a user selected instance. Therefore a concept C, called a class, in the concept hierarchy is represented by an abstract concept description and a list of pointers to each child concept $M(C)=\{C_1, C_2, ..., C_i, ..., C_n\}$, where C_i is the child concept, called subclass of concept C.

Our conceptual clustering algorithm presented here is based on similarities, because we do not consider logical but numerical concepts [19].

The output of our algorithm for applying eight exemplary shape cases of fungal strain *Ulocladium Botrytis* is shown in Figure 5. On top level the root node is shown which comprises the set of all input cases. Successively the tree is partitioned into nodes until each input case forms its own cluster.

The main advantage of our conceptual clustering algorithm is that it brings along a concept description. Thus, in comparison to agglomerative clustering methods, it is easy to understand why a set of cases forms a cluster. During the clustering process the representative case, and also the variances and maximum distances in relation to this representative case, are calculated, since they are part of the concept description. The algorithm is of incremental fashion, because it is possible to incorporate new cases into the existing learnt hierarchy.

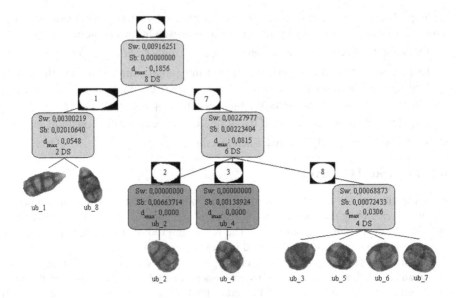

Fig. 5. Output of the Conceptual Clustering Algorithm for 2-D Shapes obtained from Fungal Spores

7 Results

The kinds of cells that are considered in this application are HEp-2 cells, which are used for the identification of antinuclear autoantibodies (ANA). ANA testing for the assessment of systemic and organ-specific autoimmune diseases has increased progressively since immunofluorescence techniques were first used to demonstrate antinuclear antibodies in 1957. HEp-2 cells allow for recognition of over 30 different nuclear and cytoplasmic patterns, which are given by upwards of 100 different autoantibodies.

The identification of the patterns has up to now been done manually by a human inspecting the slides with the help of a microscope. The lacking automation of this technique has resulted in the development of alternative techniques based on chemical reactions, which do not have the discrimination power of the ANA testing. An automatic system would pave the way for a wider use of ANA testing.

Prototypical images of HEp-2 cell patterns for six different classes are shown in Figure 6. The images were taken by an image-acquisition unit consisting of a microscope AXIOSKOP from Carl Zeiss Jena, coupled with a video camera.

In a knowledge-acquisition process [25] with a human operator, using an interview technique and a repertory grid method, we acquired the knowledge of this operator, while classifying the different cell types. Some of this knowledge is shown in table 1. The symbolic terms show that a mixture of different image information is necessary for classification. The operator uses the intensity as well as some texture information. In addition, the appearances of the cell parts within the cells are of importance, like "dark nucleoi", which also requires spatial information.

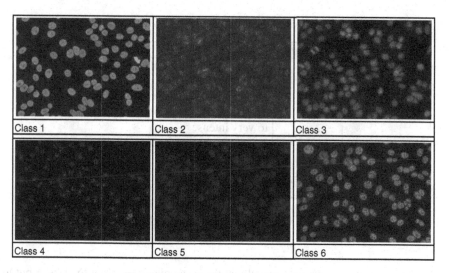

Fig. 6. Prototypical Images of Six Classes

Each image is processed by the image-analysis procedure described in the previous section. The color image is transformed into a gray-level image. The image is normalized to the mean and standard gray level calculated from all images to avoid invariance caused by the inter-slice staining variations. Automatic thresholding has been performed by the algorithm described in Section 4.1. For the objects in each slice, features based on the texture descriptor described in Section 5 are calculated for classification [26]. The first one is a simple Boolean feature which expresses the occurrence or non-occurrence of objects in the slice image. Then the number of objects in the slice image is calculated. From the objects, the area, a shape factor, and the length of the contour are calculated. The mean value for each feature is calculated over all the objects in the slice image. This is done in order to reduce the dimension of the feature vector. Since the quantization of the gray level was done in equal steps and without considering the real nature, we also calculated for each class the mean value of the gray level and the variance of the gray level. A total of 192 features were calculated that make up a very intelligent structure and texture descriptor for cells [26]. The data base created from 7-10 images per class which made up 200 cells per class is given to our decision tree unit. This unit learns the classification knowledge based on decision tree induction. Finally, the system was evaluated based on cross validation. The final result is shown in table 2. The overall classification accuracy is 92.73%. The class specific classification accuracy [1] is shown for each class in table 2 on the right side of the table and the classification quality for each class in the bottom line of the table. In most of the classes we achieved good classification accuracy. There are only few classes where the classification accuracy is not as good as the other ones. It is interesting to note that in case of class_5 four cases got misclassified as class_14 "U1-RNP" but when checking with the expert it tended out that the classifier put these samples in the right class. The case was that the expert mislabeled the cases as class_5

Table 1. Some knowledge about the class description given by a human operator

Class	ClassName	Description
Homogeneous nuclei fluorescence	Class_1	Smooth and uniform fluorescence of the nuclei. Nuclei appear sometimes dark. The chromosome fluorescence is from weak to very intense
Fine speckled nuclei fluorescence	Class_2	Dense fine speckled fluorescence
...	...	...
Nuclei fluorescence	Class_9	Nuclei are weakly homogenous or fine-grained and can hardly be discerned from the background

while the automatic system recognized that these samples belong not to class_5 but to class_14. This example shows nicely that an automatic system can lead to standardization of cell image classification. It provides objective results, it works constantly without getting tired and the results are reproducible.

The computation time of an image for the Hep-2 application is 10 seconds by an image size of 1600x1200. This computation time is fast enough for the considered application and for most other applications. Users who like to have a faster computation time can easily speed up the computation time by parallelization. Parallelization can be done in the simplest case by using more than one computer. In the hardest case, the whole algorithm can be programmed in parallel fashion.

The methods developed within the framework *Cell_Interpret* have been applied to many different applications of microscopic cell images including Hep-2 cell, Hela-cells and Malaria diagnosis. They showed to be flexible enough for different kind of cell images diagnosis tasks and they efficiently enabled the mining of the relevant knowledge for the development of an automatic image interpretation system.

The Hep-PAD version developed based on *Cell_Interpret* has been licensed to qualified industries and is meanwhile a commercial application in usage at different medical laboratories.

We are currently further developing the framework of *Cell_Interpret* to video microscopy and developing more feature extraction and image mining procedure that can further support the image mining process.

8 Expert Opinion

Recent developments are highly application oriented. Often the system works only in a semi-automatic modus [27][28] that puts a lot of work to the user using the system. Standard image processing methods are applied to specific tasks combined with a lot of heuristics [31] to make the methods more or less automatically work on the specific images.

Table 2. Results for Hep-2 Pattern Analysis

Example: Result LDS6 and DIM4

	1 AmaCent	2 Actin	3 AMA Who	4 Centromer	5 CoarseSp	6 Homogen	7 Jo-1	8 Nucleolaer	9 PMSCL	10 SCL70	11 Speckled	12 SS-A	13 SS-B	14 U1-RNP	15 Vimentin	Sum	Class Specific Quality CSQ
AmaCent	6															6	100,00%
Actin		7														7	100,00%
AMA Who			7													7	100,00%
Centromer				7												7	100,00%
CoarseSp					5						2					7	71,43%
Homogen						8										8	100,00%
Jo-1							6									6	100,00%
Nucleolaer								7								7	100,00%
PMSCL									7							7	100,00%
SCL70										8						8	100,00%
Speckled											6					6	100,00%
SS-A							1					7				8	87,50%
SS-B													7			7	100,00%
U1-RNP					4						1			7		12	58,33%
Vimentin															7	7	100,00%
Sum	6	7	7	7	9	8	7	7	7	8	9	7	7	7	7	110	94,48%
Cl. Qual.	100,00%	100,00%	100,00%	100,00%	55,56%	100,00%	85,71%	100,00%	100,00%	100,00%	66,67%	100,00%	100,00%	100,00%	100,00%		

Classification Quality

Total Number of samples	110	110
Correct classied samples	102	106
Correctness	92,73%	96,36%
Error rate	7,27%	3,64%

One such method is the Watershed-Transformation (WT) for image segmentation [27-31]. We have developed a flexible and automatic Case-Based Watershed Transformation method where the WT can be adapted to the image characteristics of the image under consideration.

Standard texture feature extraction procedures are used as well [32] but the random set approach as described here does have the flexibility to describe the different particles appearing in a cell and their randomness.

Application-oriented systems that can only solve one specific task are very costly and it takes time to develop them. The success of automatic image-interpretation systems can only be guaranteed when the development effort is as low as possible and when they can be adapted quickly to different needs and tasks. The proposed architecture of *Cell_Interpret* will help to overcome this problem.

There are commercial High-Content Analysis developments where data mining capabilities are included in the system. However, a better understanding of when and how to apply these methods and how to interpret the results are necessary for the user. Therefore we are constantly working on a methodology of data mining that is presented in our data mining tutorial (www.data-mining-tutorial.de) and copied in our data mining tools included in *Cell_Interpret*.

Another interesting observation in high-content analysis is that of images are created by using different staining to make specific cell details/objects visible [33][34]. It is obvious that in the resulting images the specific object details/parts are most visible and the analysis of these images can be simply made. However for a computer vision expert arises the question if this approach is really necessary in all case studies or would it be better to consider the whole task as a pattern recognition problem as has been done in the HEp-2 cell application and study the different patterns that appear when treating the cells in different ways. This statement might be a bit provocative and we have to admit that we do not know all applications in HCA but we would be happy to further discuss this with experts from the domain.

We also think that a better categorization of the different image analysis tasks is necessary to ensure a standardization of the image analysis procedures in HCA. A first study in that direction has been given in [35] [36]. Biologists, computer scientists and all other people involved in this field need to further discuss this and find a common basis of understanding.

The case-based reasoning approach in our system architecture *Cell_Interpret* we are recently been further developing for cell-tracking and 3D image analysis.

9 Conclusion

In this paper we have presented our architecture, *Cell_Interpret,* for High-Content Image Analysis and the methods used for the different tasks such as image segmentation, feature extraction, image mining and classification and interpretation. Most of the methods are based on case-based reasoning. CBR solves problems using already stored knowledge, and captures new knowledge, making it immediately available for solving the next problem. Therefore, case-based reasoning can be seen as a method for problem solving, and also as a method to capture new experience and make it immediately available for problem solving. It can be seen as a learning and knowledge-discovery approach, since it can capture from new experience some general knowledge, such as case classes, prototypes and some higher-level concepts.

The idea of case-based reasoning originally came from the cognitive science community which discovered that people are reasoning on formerly successfully solved cases rather than on general rules. Our interest is to build intelligent flexible and robust data-interpreting systems [37][38][39] that are inspired by the human case-based reasoning process and by doing so to model the human reasoning process when interpreting the cell images.

References

1. Perner, P.: Data Mining on Multimedia Data. LNCS, vol. 2558. Springer, Heidelberg (2002)
2. Perner, P.: Utility model, computer system for the automatic data analysis, classification, interpretation and data mining of cells, cell structures, microorganism, biotic particle, parts and products in digital images, DE 20206003294 U1
3. Wang, L., Bai, J.: Threshold selection by clustering gray levels of boundary. Pattern Recognition Letters 24(12), 1983–1999 (2003)
4. Demirkaya, O., Asyali, M.H.: Determination of image bimodality thresholds for different intensity distributions. Signal Processing: Image Communication 19(6), 507–516 (2004)
5. Patricio, M.A., Maravall, D.: A novel generalization of the gray-scale histogram and its application to the automated visual measurement and inspection of wooden Pallets. Image and Vision Computing, 2006 25(6), 805–816 (2007)
6. Pauwels, E.J., Frederix, G.: Finding Salient Regions in Images: Nonparametric Clustering for Image Segmentation and Grouping. Computer Vision and Image Understanding 75 (1-2), 73–85 (1999)
7. Cutrona, J., Bonnet, N., Herbin, M., Hofer, F.: Advances in the segmentation of multi-component microanalytical images. Ultramicroscopy 103(2), 141–152 (2005)
8. Filin, S., Pfeifer, N.: Segmentation of airborne laser scanning data using a slope adaptive neighborhood. ISPRS Journal of Photogrammetry and Remote Sensing 60(2), 71–80 (2006)
9. Kermad, C.D., Chehdi, K.: Automatic image segmentation system through iterative edge–region co-operation. Image and Vision Computing 20(8), 541–555 (2002)
10. Muñoz, X., Freixenet, J., Cufí, X., Martí, J.: Strategies for image segmentation combining region and boundary information. Pattern Recognition Letters 24(1-3), 375–392 (2003)
11. Voss, T.C., Demarco, I.A., Day, R.N.: Quantitative Imaging of Protein Interactions in the cell nucleus. Biotechniques 38(3), 413–424 (2005)
12. Beucher, S., Meyer, F.: The morphological approach of segmentation: the watershed transformation. In: Dougherty, E. (ed.) Mathematical Morphology in Image Processing, pp. 433–481. Marcel Dekker, New York (1993)
13. Perner, P.: An architecture for a CBR image segmentation system. Journal of Engineering Application in Artificial Intelligence 12(6), 749–759 (1999)
14. Frucci, M., Perner, P., Sanniti di Baja, G.: Case-based Reasoning for Image Segmentation by Watershed Transformation. In: Perner, P. (ed.) Case-Based Reasoning on Signals and Images. Springer, Heidelberg (2007)

15. Grimnes, M., Aamodt, A.: A two layer case-based reasoning architecture for medical image understanding. In: Smith, I., Faltings, B.V. (eds.) EWCBR 1996. LNCS, vol. 1168, pp. 164–178. Springer, Heidelberg (1996)

16. Knowles, D.W., Sudar, D., Bator-Kelly, C., Bissell, M.J., Lelievre, S.A.: Automated local bright feature image analysis of nuclear protein distribution identifies changes in tissue phenotype. PNAS 103(12), 445–4445 (2006)

17. Perner, P., Perner, H., Jänichen, S.: Recognition of Airborne Fungi Spores in Digital Microscopic Images. Journal Artificial Intelligence in Medicine AIM, Special Issue on CBR 36(2), 137–157 (2006)

18. Dryden, I.L., Mardia, K.V.: Statistical Shape Analysis. John Wiley & Sons, Chichester (1998)

19. Jaenichen, S., Perner, P.: Conceptual Clustering and Case Generalization of two-dimensional Forms. Computational Intelligence 22(3/4), 178–193 (2006)

20. Zamperoni, P.: Feature Extraction. In: Maitre, H., Zinn-Justin, J. (eds.) Progress in Picture Processing, pp. 121–184. Elsevier Science, Amsterdam (1996)

21. Perner, P., Perner, H., Müller, B.: Mining Knowledge for Hep-2 Cell Image Classification. Journal Artificial Intelligence in Medicine 26, 161–173 (2002)

22. Perner, P.: Prototype-Based Classification Applied Intelligence (to appear) (online available)

23. Chang, C.-L.: Finding Prototypes for Nearest Neighbor Classifiers. IEEE Trans. on Computers C-23(11), 1179–1184 (1974)

24. Wettschereck, D., Aha, D.W.: Weighting Features. In: Aamodt, A., Veloso, M.M. (eds.) ICCBR 1995. LNCS, vol. 1010, pp. 347–358. Springer, Heidelberg (1995)

25. Perner, P.: Image Mining: Issues, framework, a generic tool and its application to medical-image diagnosis. Journal Engineering Applications of Artificial Intelligence 15(2), 193–203

26. Perner, P., Perner, H., Müller, B.: Texture Classification based on Random Sets and its Application to Hep-2 Cells. In: Kasturi, R., Laurendeau, D., Suen, C. (eds.) ICPR 2002, vol. II, pp. 406–411. IEEE Computer Society, Los Alamitos (2002)

27. Gokay, K.E., Wilson, J.M.: Targeting of an Apical Endosomal Protein to Endosomes in Madin–Darby Canine Kidney Cells Requires Two Sorting Motifs. Traffic 1, 354–365 (2000)

28. Beil, M., Dürschmied, D., Paschke, St., Schreiner, B., Nolte, U., Bruel, A., Irinopoulou, T.: Spatial Distribution Patterns of Interphase Centromeres During Retinoic Acid-Induced Differentiation of Promyelocytic Leukemia Cells. Cytometry 47, 217–225 (2002)

29. Velliste, M., Murphy, R.F.: Automated determination of protein subcellular locations from 3D fluorescence microscope images. In: Proc. Biomedical Imaging 2002, pp. 867–870. IEEE Press, Los Alamitos (2002)

30. Irinopoulou, T., Vassy, J., Beil, M., Nicol, P.: Three-Dimensional DNA Image Cytometry by Confocal Scanning Laser Microscopy in Thick Tissue Blocks of Prostatic Lesions. Cytometry 27, 99–105 (1997)

31. Swedlow, J.R., Goldberg, I., Brauner, E., Sorger, P.K.: Informatics and Quantitative Analysis in Biological Imaging. Science 300(5616), 100–102 (2003)

32. Tran, D., Pham, T., Zhou, X.: Cell Phase Indentification using Fuzzy Gaussian Mixture Models. In: ISPACS 2005, International Symposium on Intelligent Signal Processing and Communication Systems, Hong Kong, China, December 14-17, 2005, pp. 465–468 (2005)

33. Lieb, J.D., Ortiz de Solorzano, C., Garcia Rodriguez, E., Jones, A., Angelo, M., Lockett, S., Meyer, B.J.: The Caenorhabditis elegans Dosage Compensation Machinery Is Recruited to X Chromosome DNA Attached to an Autosome. Genetics 156, 1603–1621 (2000)

34. Ecker, R.C., Steiner, G.E.: Microscopy-Based Multicolor Tissue Cytometry at the Single-Cell Level. Cytometry Part A 59A, 182–190 (2004)
35. Swedlow, J.R., Goldberg, I., Brauner, E., Sorger, P.K.: Informatics and Quantitative Analysis in Biological Imaging. Science 300(5616), 100–102 (2003)
36. Berlage, T.: Analyzing and mining image databases. DDT 10(11), 795–802 (2005)
37. Perner, P., Holt, A., Richter, M.: Image Processing in Case-Based Reasoning. The Knowledge Engineering Review 20(3), 311–331
38. De Mantaras, R.L., Cunningham, P., Perner, P.: Emergent case-based reasoning applications. The Knowledge Engineering Review 20(3), 325–328
39. Holt, A., Bichindaritz, I., Schmidt, R., Perner, P.: Medical applications in case-based reasoning. The Knowledge Engineering Review 20(3), 289–292

Automatic Segmentation of Unstained Living Cells in Bright-Field Microscope Images

M. Tscherepanow[1], F. Zöllner[2], M. Hillebrand[1], and F. Kummert[1]

[1] Applied Computer Science, Faculty of Technology
Bielefeld University, P.O. Box 100 131
D-33501 Bielefeld, Germany
{marko,mhillebr,franz}@techfak.uni-bielefeld.de
[2] Computer Assisted Clinical Medicine, Faculty of Medicine Mannheim
University of Heidelberg, Theodor-Kutzer-Ufer 1-3
D-68167 Mannheim, Germany
frank.zoellner@medma.uni-heidelberg.de

Abstract. The automatic subcellular localisation of proteins in living cells is a critical step in determining their function. The evaluation of fluorescence images constitutes a common method of localising these proteins. For this, additional knowledge about the position of the considered cells within an image is required. In an automated system, it is advantageous to recognise these cells in bright-field microscope images taken in parallel with the regarded fluorescence micrographs. Unfortunately, currently available cell recognition methods are only of limited use within the context of protein localisation, since they frequently require microscopy techniques that enable images of higher contrast (e.g. phase contrast microscopy or additional dyes) or can only be employed with too low magnifications. Therefore, this article introduces a novel approach to the robust automatic recognition of unstained living cells in bright-field microscope images. Here, the focus is on the automatic segmentation of cells.

1 Introduction

The complete nucleotide sequences of the genomes of a variety of species have been determined in recent years. But although we have read the genetic message of these organisms, we still do not know its meaning. Based on hereditary information, macromolecules are formed, the majority of which consists of proteins. These proteins are responsible for performing numerous functions such as assembling biological structures and controlling chemical reactions. Knowledge about their functions could enable new insights into cellular processes or facilitate the development of efficient drugs.

A common approach to determining the function of proteins is the analysis of subcellular location patterns in fluorescence microscope images [1,2,3,4,5]. Based on its location within a considered cell, conclusions about a protein's function can be drawn.

P. Perner and O. Salvetti (Eds.): MDA 2008, LNAI 5108, pp. 158–172, 2008.

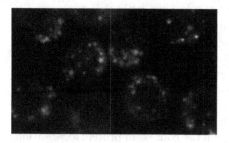

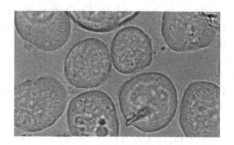

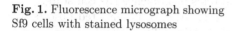

Fig. 1. Fluorescence micrograph showing Sf9 cells with stained lysosomes

Fig. 2. Bright-field image taken simultaneously with the micrograph of Fig. 1

In order to localise them, the considered proteins are tagged with a fluorescence dye, for instance with the green fluorescent protein (GFP) or one of its spectral variants [6]. Unfortunately, the surrounding cells themselves are almost invisible in these fluorescence images (see Fig. 1). Thus, additional information is required in order to associate fluorescent spots with specific cells. Commonly applied methods for the acquisition of this information are based on a manual segmentation [1,4] or the usage of stained cells [2,4].

In contrast, our approach enables an automatic segmentation of *Spodoptera frugiperda* cells (Sf9) without employing additional dyes. A bright-field microscope image, taken in parallel with each fluorescence image, is used for the identification of cells (see Fig. 2), which constitute the basis for the analysis of the corresponding fluorescence image.

The bright-field images are segmented by means of an active contour approach briefly outlined in [7]. After a discussion of relevant literature (see Section 2), this technique as well as required methods for the automatic determination of initial segments are described in Section 3. Section 4 proposes various enhancements that are relevant for a practical application of our approach. These methods are evaluated in Section 5. Eventually, the complete cell recognition approach is analysed and a short outlook is given in Section 6.

2   Related Work

In order to account for the limitations, which have to be considered within the context of automatic protein localisation in living cells, Section 2.1 introduces and evaluates basic microscopy techniques frequently used in conjunction with cell recognition approaches. As the choice for a recognition method strongly depends on the utilised microscopy technique, the application of several well-known approaches, which are discussed in Section 2.2, is partly impeded.

2.1   Applied Microscopy Techniques

A large number of cell recognition approaches such as [8,9,10] employ phase contrast microscopy to increase the contrast of acquired images. It visualises the phase shift induced by the interaction of rays of light with objects varying in thickness or refractive index. Since this microscopy technique requires special objectives that reduce the amplitude of incident light, the light from fluorescent objects would be attenuated as well. An alternation of the objective between the acquisition of the images used for protein localisation and cell recognition causes further problems, since it modifies the optical path. Consequently, an association of corresponding pixels of these images would be hampered.

Besides phase contrast microscopy, numerous approaches resort to additional dyes [2,11,12,13]. If such dyes were used within the context of protein localisation, they might interfere with the examined proteins or influence the cell state.

Bright-field microscopy, i.e. the direct observation of illuminated objects, is a widely used method for cell observation. It is usually available without any special devices. But the resulting contrast is rather low, which necessitates more complex recognition techniques [14,15,16,17]. On the other hand, bright-field microscopy is compatible with fluorescence microscopy and is probably the most frequently applied microscopy technique. Therefore, we have decided to use bright-field images as the basis of our cell recognition method.

2.2   Known Approaches to Cell Recognition

The most common approach to cell recognition consists in thresholding [18,19]. But it is often applied to nuclei rather than whole cells [20,21]. As each cell usually has a single nucleus, which covers the major fraction of its volume, these tasks are roughly equivalent.

Thresholding requires a uniform and unambiguous distribution of pixel intensities, which does not occur in bright-field images that show a great variety of cell appearances. Even if fluorescence images of stained nuclei are to be analysed, fuzzy transitions between objects and the image background may result in difficulties in selecting a proper threshold. In addition, thresholding causes problems in separating adjoining objects, which have to be dealt with separately. Here for example, the distance transform and the watershed transform can be applied [20]. Nevertheless, the prior binarisation of the image leads to a loss of information, which might be crucial for the determination of the objects' exact boundaries.

As an alternative to thresholding, there are approaches that determine and link the edges of stained nuclei using geometrical constraints [12]. Unfortunately, these constraints do not necessarily reflect the shape of visible objects – especially if these objects partially overlap.

Since subcellular structures are to be analysed after cell recognition, a high magnification (60×) is required. So, the considered cells comprise about 10,000–80,000 pixels. Therefore, methods utilising small rectangular patches in order to detect whole cells (cf. [11,13,14,15]) cannot be employed, as the computational costs would be too high. So, for example, the approach proposed in [14] takes 1 to 8 minutes to recognise cells in relatively small images (640×480) using a patch size of 625 pixels on an Intel Pentium 4 processor operating at 1.6GHz.

However, Petra Perner and her co-researchers proposed a technique for recognising fungal spores using bright-field microscopy [22], which resorts to image pyramids in order to decrease the processing time. The suggested technique iteratively compares small image regions with a set of examples for the objects under consideration, referred to as cases. Since these cases constitute images themselves, the translation, rotation and scaling of the cells have to be dealt with explicitly. A more abstract representation, for instance by means of representative features, could circumvent these problems. Furthermore, it might allow for a better generalisation, since irrelevant information can be neglected. Nevertheless, the given recognition rates appear very promising.

Cells in bright-field microscope images are separated from other cells and the surrounding by their membrane. Consequently, it is beneficial to include information about it in the segmentation procedure. This can be accomplished by determining cell membrane pixels and linking them [23,24]. But, in the case of images containing numerous cells of varying shape or size, it is difficult to obtain unambiguous solutions.

As an alternative to edge-linking methods, snakes have proven advantageous [9,10]. Besides exploiting gradient and image information, they allow for the incorporation of prior knowledge on cell features such as curvature and size without assuming a rigid model. Therefore, we decided to develop a snake-based algorithm for the recognition of Sf9 cells in bright-field images [7,16]. The component of our recognition system that performs an automatic segmentation of the images is detailed in the present article.

3   Cell Segmentation in Bright-Field Images

In order to find cells in microscope images in an automated way, several tasks are fulfilled (cf. Fig. 3). First, possible cells must be localised, i.e. the positions of candidate cells are to be determined. This enables the analysis of specific image regions instead of iteratively moving a region of interest over the complete micrograph. Here, several intermediate images are computed: one image depicting the image background (see Section 3.1) and another one showing possible cell membranes (see Section 3.2). Based on these two images, small regions within the possible cells are determined – the cell markers (see Section 3.3). They reflect the positions of the surrounding cells. Unfortunately, at this step no differentiation between real cells and other image objects is possible, since not enough information about the corresponding image objects is available.

After the localisation of candidate cells, they are segmented; that is, all pixels showing a specific cell are associated with it (see Section 3.4). Then representative features describing a cell are computed and non-cell objects can be rejected. This is achieved by means of a classifier. The classification process is introduced in [16].

3.1   Separation of Image Foreground and Background

Kenong Wu and his colleagues have shown that the local intensity variation is a valuable feature for the separation of the foreground and the background in bright-field images [17]. Instead of computing the local variation defined by the

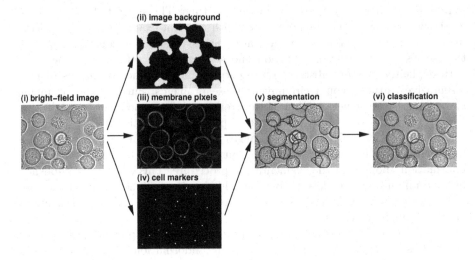

Fig. 3. Outline of the proposed cell recognition approach. On the basis of an acquired bright-field image (i) three further images which contain background pixels (ii), probable cell membrane pixels (iii) and cell markers (iv) are generated. They constitute the foundation of the proposed segmentation procedure. The segmentation (v) is followed by a classification step (vi) rejecting non-cell segments.

variance within a square neighbourhood, we take advantage of a morphological operator: the self-complementary top-hat $\varrho_S(I)$ [25].

$$\varrho_S(I) = \phi_S(I) - \gamma_S(I) \tag{1}$$

It constitutes the difference between a closing $\phi_S(I)$ and an opening $\gamma_S(I)$, which have been applied to an image I. The required structuring element is denoted by S. This operator preserves bright as well as dark image structures that cannot include S.

Figure 4 depicts the result of the application of the self-complementary top-hat to an exemplary bright-field image as well as the corresponding variance map using a square structuring element and neighbourhood of 41×41 pixels, respectively (suggested by Wu et al.).

The bimodal distribution of the local intensity variations resulting from the application of the self-complementary top-hat is considerably more distinctive than the one computed by analysing the variance. Hence, the automatic separation of image foreground and background is alleviated. Here, minimum error thresholding [26] is utilised, as it yields excellent results for the emerging grey-level distributions [17].

In order to increase the computational efficiency, structuring elements comprising 25×25 pixels have been employed. Despite their reduced size, they still perform better than the variance map using a neighbourhood of 41×41 pixels.

In principle, the application of structuring elements that do not have a rectangular shape would be possible, as well; but, since rectangular structuring elements

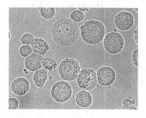 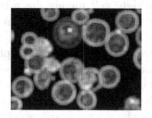 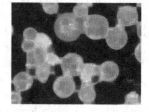

Fig. 4. Local intensity variations in a bright-field image (left). The result of the self-complementary top-hat (right) allows a noticeably better recognition of the image foreground than the variance map (centre).

can be decomposed into two linear elements, they are more computationally efficient [25]. Furthermore, other shapes do not yield considerably improved results.

3.2 Detection of Probable Cell Membrane Pixels

Probable cell membrane pixels are determined by utilising morphological operators, as well, since they enable the inclusion of knowledge concerning the shape of the image structures in question. As the cell membrane possesses a linear shape that is less curved than other cell compartments, linear structuring elements are applied. The membrane is further characterised by a substantial change of intensities between neighbouring pixels. Therefore, the gradient magnitude image is utilised instead of the original image. All image structures that cannot contain the linear structuring element – e.g. dirt, noise and intracellular objects – are removed by a morphological opening. In order to get closed contours, this operation is repeated for seven additional orientations. The resulting images are fused by computing the point-wise maximum (see Fig. 5). The whole operation constitutes an algebraic opening [25].

The length l of the linear structuring elements is crucial to the result of the algebraic opening. If it is chosen too small, irrelevant image structures will

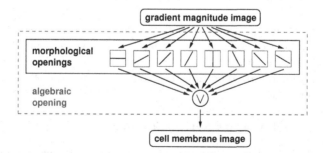

Fig. 5. Detection of pixels probably representing cell membranes. Morphological openings with linear structuring elements having eight different orientations are performed to suppress image structures that do not represent cell membranes. The resulting images are fused by a point-wise maximum operation denoted by 'V'.

remain; if the value is too high, cell membrane pixels will disappear. Hence, a procedure for the automatic determination of an optimal value had to be developed (see Section 4.1).

In order to decrease the computational effort, an optimised technique enabling the computation of morphological openings using line elements at arbitrary angles was implemented. It is based on methods proposed by Pierre Soille and his colleagues [27] who generalised an algorithm originally introduced by Marcel van Herk which solely allowed for the usage of horizontal, vertical and diagonal lines [28]. Soille's algorithm performs morphological operations independently of the length of the linear structuring elements used. In particular, images of a specific size can be processed in constant time with respect to the structuring elements' lengths.

3.3   Determination of Cell Markers

On the basis of the computed image background and cell membrane pixels, small regions within probable cells are identified – the cell markers (see Fig. 6). It is assumed that points possessing a great distance to the image background and membrane pixels lie inside cells. These points are determined by computing the local maxima of the distance transform [25].

Fig. 6. Computation of cell markers. The cell markers (right) are determined in such a way that they maximise the distance to the image background (left) and membrane pixels (centre).

In order to obtain an appropriate initialisation for the segmentation step, these regions are dilated by a small circular structuring element (diameter: 5% of the maximal cell radius, 9 pixels). Afterwards, the contours are traced so as to obtain a polygonal representation that comprises only the start and the end point of adjoining lines.

3.4   Cell Segmentation by Active Contours

Active contours have several advantages with respect to the segmentation of cells. Firstly, they always yield closed contours even if the corresponding cell membrane is barely visible. Secondly, they enable the inclusion of context-specific knowledge such as membrane curvature and cell size. So, the robustness can be improved.

Several approaches have been proposed for the computation of active contours, e.g. variational calculus, dynamic programming and greedy methods. We have decided to apply a greedy approach [29] due to its efficient computability, stability and flexibility. Since our approach aims at complete independence from user interactions while processing images, special requirements have to be fulfilled. In particular, the determined cell markers instead of close approximations of the resulting contour should be applied as initialisations.

Cohen [30] proposed a method to realise the growth of snakes by introducing an inflation force. This technique resorts to normal vectors of the contour in order to determine the direction of extension. As a result, the contour might overlap with itself if it is initialised with a concave cell marker. Hence, we have decided to utilise an alternative basis for the growth of the contours – the minimal distance to the respective initial contour. Equation (2) shows the corresponding energy functional E_{snake}^* of a parametric curve $v(x(s), y(s))$ with arc length s.

$$E_{\text{snake}}^* = \int_0^1 \left[\alpha E_{\text{cont}} + \beta E_{\text{curv}} + \gamma(E_{\text{dist}}) E_{\text{ao}} + \delta(E_{\text{dist}}) E_{\text{dist}} \right] ds \qquad (2)$$

E_{cont} and E_{curv} control the continuity and curvature, respectively. Moreover, E_{cont} fosters equal spacing between points [29]. E_{ao} represents the resulting image of the algebraic opening (see Section 3.2) and E_{dist} the distance from the initial contour. As the energies are minimised, the image as well as the distance have to be inverted. Thus, a maximal considered distance Δ_{max} is required. We have set it to the maximal cell radius increased by a tolerance interval of 20% (198 pixels in total).

The parameters α, β, γ and δ control the influence of the respective energy terms. Here, γ and δ are modified dependent on E_{dist}.

$$\gamma(E_{\text{dist}}) = \gamma_0 \cdot \frac{\Delta_{\text{max}} - E_{\text{dist}}}{\Delta_{\text{max}}} \qquad (3)$$

$$\delta(E_{\text{dist}}) = \delta_0 + \gamma_0 - \gamma(E_{\text{dist}}) \qquad (4)$$

According to Equation (3), $\gamma(E_{\text{dist}})$ yields high values if E_{dist} is small, i.e. if the snake has a great distance to its initialisation. By this, high pixel values near the cell markers, within the cells are suppressed. Equation (4) ensures that the sum of $\gamma(E_{\text{dist}})$ and $\delta(E_{\text{dist}})$ equals the sum of its base values γ_0 and δ_0, respectively. So, the extending force is reduced if the snake reaches a distance from its cell marker where the probability of membrane pixels is high. Additionally, background pixels receive a high value of E_{dist} in order to avoid an extension of the snake in this region.

This method allows for the snakes to be initialised by means of the cell markers, which are determined automatically and do barely resemble the final snakes. User interactions are not necessary.

4   Enhancements

In the introduction of our segmentation approach in Section 3 several questions were left open although they are crucial for the correct function. They are topics of current research and are answered in the following. Section 4.1 outlines a method that enables the automatic determination of the optimal length l for the linear structuring elements which are applied during the algebraic opening. A further problem consists in the parametrisation of the snakes. As they are growing, new points have to be inserted (see Section 4.2).

4.1   Optimal Length of the Linear Structuring Elements

The basis for the automatic determination of the length l of the linear structuring elements consists in n cell masks, which were manually extracted by biological experts. Besides the mask of a cell i itself, the points of a tube with a diameter of 5% of the mean cell radius that is centred at the mask's boundary are considered in order to detect the intensities of membrane pixels. The sets of the corresponding points p are denoted by $\mathcal{M}_i$ (mask) and $\mathcal{T}_i$ (tube), respectively. According to Equation (7), an optimal value l_{opt} for the length of the line elements is then computed by iterating over all possible values up to Δ_{max}.

$$I_l^{\mathcal{T}} = \sum_{i=1}^{n} \sum_{\forall p \in \mathcal{T}_i} I_l(x_p, y_p)^2 \tag{5}$$

$$I_l^{\mathcal{M}} = \sum_{i=1}^{n} \sum_{\forall p \in \mathcal{M}_i} I_l(x_p, y_p)^2 \Delta(x_p, y_p) \tag{6}$$

$$l_{opt} = \arg \max_{\forall l} \left(\frac{I_l^{\mathcal{T}}}{\max_{\forall l} I_l^{\mathcal{T}}} - \frac{I_l^{\mathcal{M}}}{\max_{\forall l} I_l^{\mathcal{M}}} \right) \tag{7}$$

$I_l(x_p, y_p)$ constitutes the image generated by an algebraic opening with a structuring element of length l. The consideration of squared pixel values results in a reduced influence of small intensities that have less negative effects on the segmentation than high ones. Moreover, the points of the mask image are weighted by their minimal distance $\Delta(x_p, y_p)$ to the boundary. l_{opt} is optimal in a sense that it maximises the difference of the intensities (scaled to fit into the interval $[0, 1]$) within both examined image regions in order to enhance the contrast.

4.2   Insertion of New Points

The segmentation consists in extending the snakes starting from small regions within probable cells. So, the distances between adjoining points are increased and a resampling of the snake, i.e. the insertion of new points is necessary. On the other hand, too high a number of points results in an increased computational effort. Thus, some kind of compromise has to be reached. Since Sf9 cells are almost elliptically shaped, an ellipse approximation of the current snake is performed [31]. This yields the lengths of the semiminor axis b and of the semimajor axis a as well as the centre C. On the basis of these values, the approximation error ϵ occurring if the ellipse is approximated by a line segment of length λ is computed (see Fig. 7).

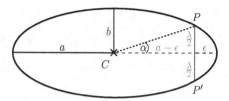

Fig. 7. Approximation of an ellipse by line segments. A line segment of length λ connecting the points P and P' causes an approximation error ϵ if it is divided equally by the major axis. As the distance between the ellipse and its centre C is maximal there, ϵ is maximal, as well. Thus, ϵ constitutes the worst case value.

An ellipse can be described by $x = a \cdot \cos \alpha$ and $y = b \cdot \sin \alpha$. Inserting the coordinates $x_P = a - \epsilon$ and $y_P = \frac{\lambda}{2}$ of point P and fusing the results leads to Equation (8) which enables the determination of λ.

$$\lambda = 2b \cdot \sin \left(\arccos \frac{a - \epsilon}{a} \right) \tag{8}$$

Instead of computing the ellipse approximation after every iteration step of the snake algorithm (variable split length, VSL), it can be applied to the determination of a constant split length λ^* (CSL). For this purpose, all manually extracted cells are approximated by an ellipse and λ^* is set to the minimal value of λ. So, a correct approximation of all cells with an error less than ϵ can be guaranteed, as well.

5   Results

We evaluated our methods on a dataset containing 499 cells manually extracted from 45 images by biological experts. In order to enable investigations regarding different foci, the dataset comprises images of the same specimens at three manually adjusted focal planes (A, B and C) showing the cell characteristics depicted in Fig. 8.

All 499 manually extracted cells were automatically marked during the preprocessing step (see Section 3.3) and each cell mask was associated with the marker closest to its centre. Furthermore, the length of the linear structuring elements for the algebraic opening was automatically set to $l_{opt} = 31$ according to Section 4.1.

In order to assess the segmentation, the manually extracted cell masks were compared with the corresponding automatically segmented cells by performing 15-fold cross-validation. The energy weights were chosen in such a way as to minimise the error term $\bar{d}^{err}$ for all except one of the images of a focal plane (see Equation (9)).

$$\bar{d}^{err} = \frac{1}{n} \sum_{i=1}^{n} \frac{d_i^{max}}{2b_i} \tag{9}$$

d_i^{max} denotes the maximal distance of corresponding manually and automatically determined contours of a cell i. These values are normed to the current manually

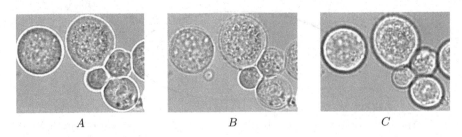

$$A \qquad\qquad B \qquad\qquad C$$

Fig. 8. Cells at different focal planes. The appearance of the examined cells varies if the focus is modified.

determined cell size represented by the length of the semiminor axis b_i of the cell's approximation by an ellipse.

After computing the energy weights, the remaining image was segmented in order to measure the test errors. $\bar{d}_A^{\text{test}}$, $\bar{d}_B^{\text{test}}$ and $\bar{d}_C^{\text{test}}$ denote the mean of these test errors over all images (see Tab. 1). Additionally, the mean point number per snake $\bar{p}$ and the average processing time[1] per image $\bar{t}$ using an AMD Athlon 64 processor (2GHz) were determined.

Table 1. Comparison of the segmentation if variable split length (VSL), constant split length (CSL) and no resampling are applied. The dash denotes parameters that were not available.

method	ϵ	λ^*	$\bar{d}_A^{\text{test}}$	$\bar{d}_B^{\text{test}}$	$\bar{d}_C^{\text{test}}$	$\bar{p}$	$\bar{t}$
VSL	0.5	–	0.104	0.118	0.142	33.9	1.038s
	0.125	–	0.088	0.109	0.139	59.2	1.200s
CSL	0.5	18	0.094	0.130	0.143	45.2	0.802s
	0.125	9	0.102	0.116	0.141	89.6	0.980s
no resampling	–	–	0.109	0.123	0.146	23.4	0.708s

The results of all methods show that the choice of the focal plane has a considerable effect on the quality of the segmentation. The errors rise from plane A to plane C. These results originate in less distinctive cell membranes (B) and stronger intracellular intensity variations (C), respectively (cf. Fig. 8)

Both reparametrisation methods attained smaller segmentation errors than the original approach which does not perform resampling. Since CSL utilises a minimal value of the split length λ that is sufficient for all cells, it requires additional points in comparison to VSL. These unnecessary additional points seem to deteriorate the segmentation compared to VSL (e.g. for $\epsilon = 0.125$). The lowest errors were reached by VSL with $\epsilon = 0.125$, which required significantly more processing time than the other methods because of the determination of λ during the actual segmentation. So, if enough time is available VSL should be employed. Otherwise, the original approach and CSL, especially with $\epsilon = 0.5$, are beneficial.

---

[1] Excluding the time for the computation of the cell markers.

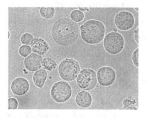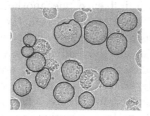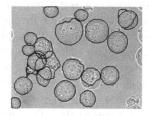

Fig. 9. Segmentation of a bright-field image (left). The final snakes are depicted as dark contours. The central image only shows segments that could be associated with manually extracted cell masks, whereas the right picture comprises all snakes.

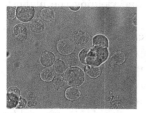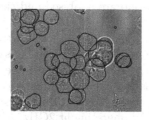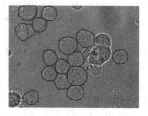

Fig. 10. Cells recognised in an independent test image (left). After the classification (right) only segments enclosing individual cells remain. Other regions yielded by the segmentation procedure (centre) are dropped.

In order to assess our results, the manually extracted segments of 363 cells determined by five people were compared pairwisely. The corresponding contours possess a mean maximal distance of 5% of the cell size with a standard deviation of 2.5%, as the cell membranes cannot always be determined unambiguously. Thus, we conclude that our methods are very accurate.

6   Conclusion

We have presented an approach to the automatic segmentation of cells in bright-field microscope images. Furthermore, several enhancements with respect to the quality of the preprocessing as well as the segmentation have been introduced. The result of our segmentation procedure is depicted by Fig. 9. Here, VSL with $\epsilon = 0.125$ was applied.

The evaluation of our segmentation method occurred based on images at three different focal planes in order to enable the choice of an optimal one. At this focal plane, all methods yielded excellent results insofar as the segmentation error is only slightly higher (difference $< 10\%$) than the deviations of segments manually determined by different persons.

As several cells receive multiple cell markers and other image objects – for instance dead cells and dirt – are marked as well, the number of segments is higher than the number of cells. But since the snakes are grown independently, overlapping segments do not influence one another. Furthermore, if the cell marker is situated in the centre of a real cell, the respective cell is usually segmented properly. In principle, this even enables a correct segmentation of overlapping cells. But as Sf9 cells grow adherently and the cell density is kept relatively low, the cells usually do not overlap in our images. If a cell marker does not lie close to a cell's centre, arbitrarily shaped segments might result. These segments have to be sorted out. Hence, a classification of the final snakes is performed [16]. The result of the application of our complete cell recognition approach to an independent test image is shown in Fig. 10.

Assuming a cell has been segmented by multiple snakes, the one best representing the cell should be chosen. Therefore, a value is computed, which reflects the confidence in the classification result [32]. Afterwards, segments that represent cells and strongly overlap are analysed and the one yielding the highest confidence value is selected.

Based on a test set comprising 302 cell masks extracted from 19 bright-field micrographs (focal planes A and B), recognition rates up to 90.1% were reached. The corresponding mean segmentation error of the recognised cells amounts to $\bar{d}^{rec}=0.11$, which confirms the results summarised by Tab. 1. The complete processing of a bright-field image (1344×1024 pixels) takes approximately 15s using an AMD Athlon 64 CPU operating at 2GHz. Although this is comparably fast, techniques such as image pyramids could probably be applied to realise a reduction of the computational load. But even without such an acceleration, this approach is suitable to be applied in conjunction with a high-throughput protein localisation technique. The first experiments regarding the usage of our cell recognition approach in conjunction with a protein localisation method have led to very promising results [5].

Recently, we have conducted experiments using an alternative cell line originating from the fruit fly *Drosophila melanogaster* [32]. Although these cells are considerably more difficult to recognise, our cell recognition method – after it had been adapted to them – performed very well.

References

1. Huh, W.K., Falvo, J.V., Gerke, L.C., Carroll, A.S., Howson, R.W., Weissman, J.S., O'Shea, E.K.: Global analysis of protein localization in budding yeast. Nature 425, 686–691 (2003)
2. Liebel, U., Starkuviene, V., Erfle, H., Simpson, J.C., Poustka, A., Wiemann, S., Pepperkok, R.: A microscope-based screening platform for large-scale functional protein analysis in intact cells. FEBS Letters 554, 394–398 (2003)
3. Murphy, R.F., Velliste, M., Porreca, G.: Robust numerical features for description and classification of subcellular location patterns in fluorescence microscope images. Journal of VLSI Signal Processing 35, 311–321 (2003)
4. Chen, X., Murphy, R.F.: Interpretation of Protein Subcellular Location Patterns in 3D Images Across Cell Types and Resolutions. In: Hochreiter, S., Wagner, R. (eds.) BIRD 2007. LNCS (LNBI), vol. 4414, pp. 328–342. Springer, Heidelberg (2007)

5. Tscherepanow, M., Kummert, F.: Subcellular localisation of proteins in living cells using a genetic algorithm and an incremental neural network. In: Bildverarbeitung für die Medizin 2007, pp. 11–15. Springer, Heidelberg (2007)

6. Tsien, R.Y.: The green fluorescent protein. Annual Review of Biochemistry 67, 509–544 (1998)

7. Tscherepanow, M., Zöllner, F., Kummert, F.: Aktive Konturen für die robuste Lokalisation von Zellen. In: Bildverarbeitung für die Medizin 2005, pp. 375–379. Springer, Heidelberg (2005)

8. Debeir, O., Ham, P.V., Kiss, R., Decaestecker, C.: Tracking of migrating cells under phase-contrast video microscopy with combined mean-shift processes. IEEE Transactions on Medical Imaging 24, 697–711 (2005)

9. Ray, N., Acton, S.T., Ley, K.: Tracking leukocytes in vivo with shape and size constrained active contours. IEEE Transactions on Medical Imaging 21, 1222–1235 (2002)

10. Zimmer, C., Labruyère, E., Meas-Yedid, V., Guillén, N., Olivo-Marin, J.C.: Segmentation and tracking of migrating cells in videomicroscopy with parametric active contours: A tool for cell-based drug testing. IEEE Transactions on Medical Imaging 21, 1212–1221 (2002)

11. Nattkemper, T.W., Wersing, H., Ritter, H., Schubert, W.: A neural network architecture for automatic segmentation of fluorescence micrographs. Neurocomputing 48, 357–367 (2002)

12. Raman, S., Maxwell, C.A., Barcellos-Hoff, M.H., Parvin, B.: Geometric approach to segmentation and protein localization in cell culture assays. Journal of Microscopy 225, 22–30 (2007)

13. Schubert, W., Friedenberger, M., Bode, M., Philipsen, L., Ritter, H., Nattkemper, T.W.: Automatic recognition of muscle invasive T-lymphocytes expressing dipeptidyl-peptidase IV (CD26), and analysis of the associated cell surface phenotypes. Journal of Theoretical Medicine 4, 67–74 (2002)

14. Long, X., Cleveland, W.L., Yao, Y.L.: Effective automatic recognition of cultured cells in bright field images using Fisher's linear discriminant preprocessing. Image and Vision Computing 23, 1203–1213 (2005)

15. Long, X., Cleveland, W.L., Yao, Y.L.: Automatic detection of unstained viable cells in bright field images using a support vector machine with an improved training procedure. Computers in Biology and Medicine 6, 339–362 (2006)

16. Tscherepanow, M., Zöllner, F., Kummert, F.: Classification of segmented regions in brightfield microscope images. In: Proceedings of the International Conference on Pattern Recognition (ICPR), vol. 3, pp. 972–975. IEEE, Los Alamitos (2006)

17. Wu, K., Gauthier, D., Levine, M.: Live cell image segmentation. IEEE Transactions on Biomedical Engineering 42, 1–12 (1995)

18. Chen, X., Yu, C.: Application of some valid methods in cell segmentation. In: Proceedings of SPIE, vol. 4550, pp. 340–344 (2001)

19. Grobe, M., Volk, H., Münzenmayer, C., Wittenberg, T.: Segmentierung von überlappenden Zellen in Fluoreszenz- und Durchlichtaufnahmen. In: Bildverarbeitung für die Medizin 2003, pp. 201–205. Springer, Heidelberg (2003)

20. Malpica, N., de Solórzano, C.O., Vaquero, J.J., Santos, A., Vallcorba, I.M., García-Sagredo, J., del Pozo, F.: Applying watershed algorithms to the segmentation of clustered nuclei. Cytometry 23, 289–297 (1997)

21. Walker, R.F., Jackway, P.T., Lovell, B.: Classification of cervical cell nuclei using morphological segmentation and textural feature extraction. In: Australian and New Zealand Conference on Intelligent Information Systems, pp. 297–301 (1994)

22. Perner, P., Jänichen, S., Perner, H.: Case-based object recognition for airborne fungi recognition. Artificial Intelligence in Medicine 36, 137–157 (2006)
23. Alexopoulos, L.G., Erickson, G.R., Guilak, F.: A method for quantifying cell size from differential interference contrast images: validation and application to osmotically stressed chondrocytes. Journal of Microscopy 205, 125–135 (2002)
24. Young, D., Gray, A.J.: Cell identification in differential interference contrast microscope images using edge detection. In: Proceedings of the 7th British Machine Vision Conference (BMVC), vol. 1, pp. 133–142. BMVA Press (1996)
25. Soille, P.: Morphological Image Analysis: Principles and Applications. Springer, New York (2003)
26. Kittler, J., Illingworth, J.: Minimum error thresholding. Pattern Recognition 19, 41–47 (1986)
27. Soille, P., Breen, E.J., Jones, R.: Recursive implementation of erosions and dilations along discrete lines at arbitrary angles. IEEE Transactions on Pattern Analysis and Machine Intelligence 18, 562–667 (1996)
28. van Herk, M.: A fast algorithm for local minimum and maximum filters on rectangular and octagonal kernels. Pattern Recognition Letters 13, 517–521 (1992)
29. Williams, D.J., Shah, M.: A fast algorithm for active contours and curvature estimation. Computer Vision, Graphics, and Image Processing: Image Understanding 55, 14–26 (1992)
30. Cohen, L.D.: Note: On active contour models and balloons. Computer Vision, Graphics, and Image Processing: Image Understanding 53, 211–218 (1991)
31. Fitzgibbon, A.W., Pilu, M., Fisher, R.B.: Direct least square fitting of ellipses. IEEE Transactions on Pattern Analysis and Machine Intelligence 21, 476–480 (1999)
32. Tscherepanow, M., Jensen, N., Kummert, F.: Recognition of unstained live Drosophila cells in microscope images. In: Proceedings of the International Machine Vision and Image Processing Conference (IMVIP), pp. 169–176. IEEE, Los Alamitos (2007)

Author Index

Lecture Notes in Artificial Intelligence (LNAI)

Vol. 4840: L. Paletta, E. Rome (Eds.), Attention in Cognitive Systems. XI, 497 pages. 2007.

Vol. 4830: M.A. Orgun, J. Thornton (Eds.), AI 2007: Advances in Artificial Intelligence. XIX, 841 pages. 2007.

Vol. 4828: M. Randall, H.A. Abbass, J. Wiles (Eds.), Progress in Artificial Life. XII, 402 pages. 2007.

Vol. 4827: A. Gelbukh, Á.F. Kuri Morales (Eds.), MICAI 2007: Advances in Artificial Intelligence. XXIV, 1234 pages. 2007.

Vol. 4826: P. Perner, O. Salvetti (Eds.), Advances in Mass Data Analysis of Signals and Images in Medicine, Biotechnology and Chemistry. X, 183 pages. 2007.

Vol. 4819: T. Washio, Z.-H. Zhou, J.Z. Huang, X. Hu, J. Li, C. Xie, J. He, D. Zou, K.-C. Li, M.M. Freire (Eds.), Emerging Technologies in Knowledge Discovery and Data Mining. XIV, 675 pages. 2007.

Vol. 4811: O. Nasraoui, M. Spiliopoulou, J. Srivastava, B. Mobasher, B. Masand (Eds.), Advances in Web Mining and Web Usage Analysis. XII, 247 pages. 2007.

Vol. 4798: Z. Zhang, J.H. Siekmann (Eds.), Knowledge Science, Engineering and Management. XVI, 669 pages. 2007.

Vol. 4795: F. Schilder, G. Katz, J. Pustejovsky (Eds.), Annotating, Extracting and Reasoning about Time and Events. VII, 141 pages. 2007.

Vol. 4790: N. Dershowitz, A. Voronkov (Eds.), Logic for Programming, Artificial Intelligence, and Reasoning. XIII, 562 pages. 2007.

Vol. 4788: D. Borrajo, L. Castillo, J.M. Corchado (Eds.), Current Topics in Artificial Intelligence. XI, 280 pages. 2007.

Vol. 4775: A. Esposito, M. Faundez-Zanuy, E. Keller, M. Marinaro (Eds.), Verbal and Nonverbal Communication Behaviours. XII, 325 pages. 2007.

Vol. 4772: H. Prade, V.S. Subrahmanian (Eds.), Scalable Uncertainty Management. X, 277 pages. 2007.

Vol. 4766: N. Maudet, S. Parsons, I. Rahwan (Eds.), Argumentation in Multi-Agent Systems. XII, 211 pages. 2007.

Vol. 4760: E. Rome, J. Hertzberg, G. Dorffner (Eds.), Towards Affordance-Based Robot Control. IX, 211 pages. 2008.

Vol. 4755: V. Corruble, M. Takeda, E. Suzuki (Eds.), Discovery Science. XI, 298 pages. 2007.

Vol. 4754: M. Hutter, R.A. Servedio, E. Takimoto (Eds.), Algorithmic Learning Theory. XI, 403 pages. 2007.

Vol. 4737: B. Berendt, A. Hotho, D. Mladenic, G. Semeraro (Eds.), From Web to Social Web: Discovering and Deploying User and Content Profiles. XI, 161 pages. 2007.

Vol. 4733: R. Basili, M.T. Pazienza (Eds.), AI*IA 2007: Artificial Intelligence and Human-Oriented Computing. XVII, 858 pages. 2007.

Vol. 4724: K. Mellouli (Ed.), Symbolic and Quantitative Approaches to Reasoning with Uncertainty. XV, 914 pages. 2007.

Vol. 4722: C. Pelachaud, J.-C. Martin, E. André, G. Chollet, K. Karpouzis, D. Pelé (Eds.), Intelligent Virtual Agents. XV, 425 pages. 2007.

Vol. 4720: B. Konev, F. Wolter (Eds.), Frontiers of Combining Systems. X, 283 pages. 2007.

Vol. 4702: J.N. Kok, J. Koronacki, R. Lopez de Mantaras, S. Matwin, D. Mladenič, A. Skowron (Eds.), Knowledge Discovery in Databases: PKDD 2007. XXIV, 640 pages. 2007.

Vol. 4701: J.N. Kok, J. Koronacki, R. Lopez de Mantaras, S. Matwin, D. Mladenič, A. Skowron (Eds.), Machine Learning: ECML 2007. XXII, 809 pages. 2007.

Vol. 4696: H.-D. Burkhard, G. Lindemann, R. Verbrugge, L.Z. Varga (Eds.), Multi-Agent Systems and Applications V. XIII, 350 pages. 2007.

Vol. 4694: B. Apolloni, R.J. Howlett, L. Jain (Eds.), Knowledge-Based Intelligent Information and Engineering Systems, Part III. XXIX, 1126 pages. 2007.

Vol. 4693: B. Apolloni, R.J. Howlett, L. Jain (Eds.), Knowledge-Based Intelligent Information and Engineering Systems, Part II. XXXII, 1380 pages. 2007.

Vol. 4692: B. Apolloni, R.J. Howlett, L. Jain (Eds.), Knowledge-Based Intelligent Information and Engineering Systems, Part I. LV, 882 pages. 2007.

Vol. 4687: P. Petta, J.P. Müller, M. Klusch, M. Georgeff (Eds.), Multiagent System Technologies. X, 207 pages. 2007.

Vol. 4682: D.-S. Huang, L. Heutte, M. Loog (Eds.), Advanced Intelligent Computing Theories and Applications. XXVII, 1373 pages. 2007.

Vol. 4676: M. Klusch, K.V. Hindriks, M.P. Papazoglou, L. Sterling (Eds.), Cooperative Information Agents XI. XI, 361 pages. 2007.

Vol. 4667: J. Hertzberg, M. Beetz, R. Englert (Eds.), KI 2007: Advances in Artificial Intelligence. IX, 516 pages. 2007.

Vol. 4660: S. Džeroski, L. Todorovski (Eds.), Computational Discovery of Scientific Knowledge. X, 327 pages. 2007.

Vol. 4659: V. Mařík, V. Vyatkin, A.W. Colombo (Eds.), Holonic and Multi-Agent Systems for Manufacturing. VIII, 456 pages. 2007.

Vol. 4651: F. Azevedo, P. Barahona, F. Fages, F. Rossi (Eds.), Recent Advances in Constraints. VIII, 185 pages. 2007.

Vol. 4648: F. Almeida e Costa, L.M. Rocha, E. Costa, I. Harvey, A. Coutinho (Eds.), Advances in Artificial Life. XVIII, 1215 pages. 2007.

Vol. 4635: B. Kokinov, D.C. Richardson, T.R. Roth-Berghofer, L. Vieu (Eds.), Modeling and Using Context. XIV, 574 pages. 2007.

Vol. 4632: R. Alhajj, H. Gao, X. Li, J. Li, O.R. Zaïane (Eds.), Advanced Data Mining and Applications. XV, 634 pages. 2007.

Vol. 4629: V. Matoušek, P. Mautner (Eds.), Text, Speech and Dialogue. XVII, 663 pages. 2007.

Vol. 4626: R.O. Weber, M.M. Richter (Eds.), Case-Based Reasoning Research and Development. XIII, 534 pages. 2007.

Vol. 4617: V. Torra, Y. Narukawa, Y. Yoshida (Eds.), Modeling Decisions for Artificial Intelligence. XII, 502 pages. 2007.